Behind Ethical Consumption

Gianluigi Guido

Behind Ethical Consumption

Purchasing motives and marketing strategies for organic food products, non-GMOs, bio-fuels

PETER LANG
Bern · Berlin · Bruxelles · Frankfurt am Main · New York · Oxford · Wien

Information bibliographique publiée par «Die Deutsche Bibliothek»
«Die Deutsche Bibliothek» répertorie cette publication dans la «Deutsche Nationalbibliografie»; les données bibliographiques détaillées sont disponibles sur Internet sous ‹http://dnb.ddb.de›.

British Library Cataloguing-in-Publication Data: A catalogue record for this book is available from The British Library, Great Britain

Library of Congress Cataloging-in-Publication Data

Guido, Gianluigi.
Behind ethical consumption : purchasing motives and marketing strategies for organic food products, non-GMOs, bio-fuels / [Gianluigi Guido].
p. cm.
Includes bibliographical references and index.
ISBN 978-3-0343-0095-7 (alk. paper)
1. Consumer behavior–Moral and ethical aspects. 2. Consumer behavior–Psychological aspects. 3. Consumption (Economics)–Moral and ethical aspects. 4. Marketing–Psychological aspects. 5. Natural foods–Social aspects. 6. Genetically modified foods–Social aspects. 7. Biomass energy–Social aspects. I. Title.
HF5415.32.G85 2009
641.3'020688--dc22

2009028930

Cover picture: Valentina D'Amaro, Senza titolo (2005), oil on canvas, courtesy of the author

Cover design: Thomas Jaberg, Peter Lang AG

ISBN 978-3-0343-0095-7

Hochfeldstrasse 32, CH-3012 Bern
info@peterlang.com, www.peterlang.com, www.peterlang.net

Printed in Germany

Contents

Tables

Figures

Gianluigi Guido

1. Ethics in consumer buying

1.1. An introduction

This volume is based on a research program that I conceived and directed, in Italy, at the University of Salento in Lecce, and both at LUISS University and at *Sapienza* University in Rome, on the topic of *motives behind ethical consumption.*

The area of ethical decision-making is vast (Cooper-Martin and Holbrook 1993), and so it is difficult to find an adequate definition for "ethical consumption". One must first define the moral base of ethical consumerism (Caruana 2007). According to Crane and Matten (2004: 11), "morality is concerned with the norms, values and beliefs embedded in social processes which define right and wrong for an individual or community." Morality constitutes the object of ethics, envisaged as a branch of philosophy, applicable to human choices. From this, it follows therefore that, when one considers consumer choices, *ethical consumption* can be defined as "the degree to which consumers prioritize their own ethical concerns when making product choices" (Shaw and Clarke 1998: 163). Specifically, "ethical consumption" can be considered a *consumer philosophy* akin to so-called "fair trade" – a market-based approach that aims to help producers in developing countries and promote sustainability – as well as "green consumerism", not to mention "voluntary simplicity" (Connolly and Shaw 2006). The latter understood as the downshifting of maximum consumption – and, possibly, income – in response to an inner moral tension which drives one to satisfy not only one's material desires but also one's spiritual urges and ethical commitments (Shaw and Newholm 2002).

Individuals engaged in ethical consumption – so-called *ethical consumers* – are socially-conscious consumers who take into account the public consequences of their private acts of consumption and who can also use their power to bring about social change (Newholm and Shaw 2007). Ethical consumers can express themselves, therefore, in various ways (Harper and Makatouni 2002) ranging from simply buying products which are not harmful to society or the environment, or which respect animal and human rights, such as free-range eggs, to more complex and committed behaviors, like boycotting goods produced by child labor.[1] They choose, buy, and evaluate products (Wheale and Hinton 2007) or consciously behave in a certain way (Clarke 2006; Shaw *et al.* 2005) which is consistent with their moral norms and their espousal of certain environmental and societal issues, such as opposition to pollution, animal testing, oppressive regimes, illegal working conditions, the arms trade, and so on. Consequently, *ethical consumer* is a term that covers a very wide spectrum of person.

This phenomenon has been increasing in intensity, as clearly emerges from a recent study conducted at an international level by Manget, Roche and Münnich (2009) at the Boston Consulting Group. The results of this survey show, for example, that 50% of consumers worldwide buy "green" products, that are eco-friendly. The origins for the rise in ethical consumerism from the end of the 20th century onwards are traceable, according to many scholars, to increased media coverage and greater access to information, as well as a better availability of such products, to a greater awareness of the expression of

---

1 *Boycott* is a typical action through which consumers deliberatively avoid purchasing products offered by firms which violate shared ethical principles. Boycotting actions are usually targeted at leading multinational companies: e.g. Nestlé in the food industry; Nike in the sportswear sector; and Shell among the petrochemical giants (e.g., Post 1985). A more generalized tactic which uses purchasing power against ethically irresponsible firms is the so-called "buycott", which consists in the adoption of sustainable consumption styles. It may express itself in the reduced purchase and consumption of products as compatible with personal moral norms (Shaw and Newholm 2002). Typical behaviors in this field are, for example, following a simplified (or a vegetarian) diet, using the car less, and buying second-hand products, as well as eating organic food products and refusing genetically modified ones.

one's own consumer behaviors, and to the desire to escape from the risks associated with consumption in an individualistic society, in those cases where traditions lose their influence as guides to behavior which may be beneficial to those cultures (cf. Newholm and Shaw 2007).

Yet, aside from these motives which relate to factors beyond the decision-makers' control, that is, which are not directly determined by each consumer, it is fitting to investigate the personal motivations that drive an individual to consume in a certain way. It is far from certain, indeed, that there exists such a thing as an "ethical consumer" *tout court* as distinct from a "traditional consumer" (cf. Carrigan and Attalla 2001): considering the wide range of activities which may be undertaken by consumers to fulfill their moral obligations, it is relevant to evaluate the intrinsic motives which may push consumers to "act ethically", whether they are acting sincerely, thus absolving their moral duties, or whether they are merely adapting, e.g. submitting to social pressures of the community in which they live.

Motivations are "the processes that cause people to behave as they do" (Solomon, Bamossy and Askegaard 1999: 91) and occur when a need – intrinsic or induced – is aroused that the consumer wishes to satisfy. When it comes to consumption – and ethical consumption, in particular – motivations behind purchase behaviors are considered in a series of connected studies by Vitell and Muncy (Muncy and Vitell 1992; Vitell and Muncy 1992, 2005), who introduce the construct of *consumer ethics*. These are defined as the moral principles and standards that guide the behaviors of individuals or groups towards the purchase, use, and disposition of goods and services.[2] Consumer ethics drives the consumer to satisfy three fundamental

2 They also assessed the construct of *consumer ethics* through a list of beliefs related to subjective moral judgments on specific consumption-related actions. These actions regarded four main dimensions describing specific patterns of behavior, namely: (*i*) actively benefiting from illegal activities (e.g., "changing price-tags on merchandise in a retail store"), (*ii*) passively benefiting (e.g., "observing someone shoplifting and ignoring it"); (*iii*) actively benefiting from deceptive but legal practices (e.g., "removing the pollution control device from an automobile in order to get better mileage"), and (*iv*) no harm/no foul activities (e.g., "using computer software or games that you did not buy").

types of need (cf. Harrison, Newholm and Shaw 2005): *control needs*, the necessity (or, at times, the illusion) of having control over everyday activities and facts of life which one dislikes or fears (e.g., the rapid advances in genetic engineering); *social integration needs*, the desire to feel part of a group, accepted according to the moral mores of the rest of society; and *authenticity needs*, namely the search for the genuine (e.g. in foods, for the heath of oneself and one's family), or the return to traditions as a guide to virtuous behaviors. According to Freestone and McGoldrick (2008), these needs, which motivate ethical consumption, are viewed in different ways by consumers according to their levels of ethical awareness, ethical concern, and ethical action, seen as three levels of commitment – in order of relevance – towards ethical themes.

1.2. The study of determinants of ethical consumption

Ethics plays an important role in human activities, in general, and consumption, in particular. Many recent studies have called for research into this field. Although there is indication that many consumers are increasingly attracted by offers which are socially and environmentally responsible, even recently it has been admitted that "little is known about the motives that drive some toward, or deter others from, higher levels of ethical concern and action in their purchasing decisions." (Freestone and McGoldrick 2008: 445). According to Belk, Devinney and Eckhardt (2005: 275), "relatively little attention has been given to how ethical considerations in consumption are thought about by consumers [… such as] worker treatment, fair-trade food, genetically modified food, environmental concerns, human rights, anti-competitive practices, racial or gender discrimination, and other such issues." Also Wheale and Hinton (2007) recognize that few investigations take into account the key ethical drivers behind the decision-making process in consumption. Vitell (2003: 33), who reviews and synthesizes the major research studies in the area of consumer ethics that have appeared since 1990 in the marketing literature, notes, conclusively, that: "relatively few studies have examined ethical issues in the marketplace from the perspective of consumer ethics, but consumers are a major participant in the business process and not considering them in ethics research can result in an incomplete understanding of that process." (see also Fukukawa 2003; Vitell and Muncy 1992, for reviews).

Market trends have shown that individuals make increasing use of ethical principles when evaluating and choosing branded products. Yet ethical consumption is still a relatively under-researched area in marketing literature and the experimental studies illustrated in this book are an attempt to fill this gap. In particular, this volume examines three critical issues in ethical consumption that motivate consumer choice in this area and which, consequently, are able to influence the marketing strategy of firms. The first area of research is the *psychological determinants of consumers' purchase intention*, that is to say, the beliefs and variables underlying them, identified within a consolidated theory of social psychology – namely, Ajzen's (1991) *Theory of planned behavior* – extended to other variables (i.e., moral norms, moral disengagement, and self-identity) to take into account the specific consumer environment under consideration, that of ethically-friendly products (such as "green" products). The *Theory of planned behavior* (Ajzen 1991) is a model that, in its basic version, posits that the behavioral intention, assumed as the best predictor of an actual behavior (such as that of buying green products), is determined by three main antecedent factors. These determinants are: (*i*) *attitude*, that is, the subjective positive (or negative) predisposition towards the behavior; (*ii*) *subjective norm*, the perception of social pressures to perform (or to avoid performing) the target behavior; and (*iii*) *perceived behavioral control*, the perception of external factors that facilitate (or inhibit) the behavior. Researchers in this field have extended this theory to the ethics-related concepts that are considered as potential antecedents of intention and behavior. Minton and Rose (1997) identify *personal norms*, which pertain to the sense of moral obligation and the self-concept (the perception of the self as a person), as further main determinants of specific ethical purchase behaviors. These findings are confirmed by Shaw and colleagues (Shaw and Shiu

2002ab, 2003; Shaw, Shiu and Clarke 2000), who find *ethical obligation* and *self-identity* to make a significant contribution to explaining consumers' purchase intentions towards ethically-charged products.

The second issue addressed by this volume also concerns the motives behind ethical consumption, this time considering *values* pursued by consumers and consumers' cross-cultural individual differences. Schwartz and Bilsky (1987: 551) define values as "concepts or beliefs about desirable end states or behaviors that transcend specific situations, guide selection or evaluation of behavior or events, and are ordered by relative importance." Freestone and McGoldrick (2008: 447) drawing from Feather (1992) define *values* as "a set of stable, general beliefs about what is desirable and [they postulate…] that these emerge both from society's norms and from the individual's core psychology needs and sense of self." Studies within this second research stream have related ethical purchase behavior to *personal values*, usually measured through Schwartz's scale (Schwartz and Bilsky 1990), to identify common values that act as guiding principles for ethical consumers.[3] Values do not relate to intention and behavior directly; rather, they impact on these variables only indirectly, through more concrete and domain-specific constructs like attitude towards a specific ethically-relevant product/behavior (Follows and Jobber 2000; Honkanen, Verplanken and Olsen 2006). Moreover, the validity of this Value-Attitude-Intention causal hierarchy has been documented in different contexts and cultures. Notwithstanding this, one should bear in mind that individual differences can be found both between consumers of the same nationality (Carrigan and Attalla 2001) – one should just consider, for example, how ethical values are handed down in families and the effect that they have on each new

3 Muncy and Eastman (1998) show that such values are usually negatively correlated to the level of materialism displayed in subjects who are less interested in ethical issues – where, by *material* one refers to the degree of one's attachment to goods and earthly possessions in general (Belk 1985). By contrast, in a recent study, Vitell, Paolillo and Singh (2006) find that most consumers' ethical beliefs and behaviors are significantly explained by individual factors like *intrinsic religiosity* – i.e. the extent to which one is influenced by religion and religious beliefs – and *money ethics*, the degree to which one ascribes ethically positive (or negative) meanings to money.

generation (Carey, Shaw and Shiu 2008) – and between consumers of different nationalities (Polonsky *et al.* 2001). It is important, therefore, to verify the degree of cross-cultural homogeneity in the results obtained in trans-national studies; to gauge the size of differences between nations in ethical terms, one just has to read, for instance, the study carried out in parallel in the USA and Egypt by Al-Khatib, Vitell and Rawwas (1997), which may serve as a warning even in those cultural contexts which might, at first sight, appear more homogenous, as in the case of the European Union.

The third and final area of research covered by this work is the *latent dimensions of product image* (in particular, product personality) and how these bear upon the determinants of ethical consumers' intention to buy. *Product personality* is used here as a construct proxy of product image, because it is considered by many scholars as a more stable variable over time (cf. Dobni and Zinkhan 1990; Poiesz 1989) and, thus, better suited to describe the dimensions determined by consumers' cognitive representations, especially when considering attributes that are not only descriptive of human personality and traits that are not uniformly positive (D'Astous and Lévesque 2003). The *Prospect method* (Caprara and Barbaranelli 2000; Caprara, Barbaranelli and Guido 2000) was employed to evaluate the incidence, on the intention to buy, of these "latent" traits – as inferable, that is, from perceived attributes – which form the ethical consumers' image of the products. This model combines Ajzen's (1991) *Theory of planned behavior* and the *Five-factor model* of personality traits, originally developed in social psychology (cf. Digman 1990; Goldberg 1993, for reviews), according to which, attributes describing personality can be synthesized in five principal factors (i.e., *Big Five factors*), which summarize the image constructed by the perceivers, namely: (*i*) *Agreeableness*, the orientation towards compassion and caring about others; (*ii*) *Conscientiousness*, the preference for goal-oriented activity; (*iii*) *Emotional Stability/Neuroticism*, the ability/inability to cope effectively with negative emotions; (*iv*) *Extroversion/Introversion*, the preference/aversion for social interaction; and (*v*) *Openness to Experience*, the tolerance of new ideas and new ways of doing things. Recent research suggests that, under certain limitations, human personality may serve as a metaphor for the description of non-human entities such as brands, products and services (e.g., Caprara, Barbaranelli, and Guido 2001). Thus, the incidence that different dimensions of a brand/product's personality have on consumers' intention to buy it can be measured, provided that there is a model which establishes a theoretical framework for the determinants of purchase intention.

1.2.1. The articulation of the book

This work is divided into five separate, yet related studies, each of which deals with the issue of the motivations behind ethical choices of consumption and discusses their implications on marketing strategy. The fields of investigation selected range from organic food to genetically modified products, from bio-fuels to new low-emission transport technologies, the consumption of each of which has by its very nature a recognized ethical validity. On these themes, this volume offers a European point of view and, in particular, an Italian one, either duplicating studies undertaken in various countries and comparing the results with those found in Italy, or proposing new and original lines of research into the antecedents of purchase intentions that have never before been explored. This research was partly financed with ordinary research funds (previously called 60%) by the Italian Ministry of Instruction, Universities and Research (MIUR), and undertaken by a team of young researchers and doctoral students, under my direction.

The study reported in Chapter 2 analyzes the influence of *moral norms* and *self-identity* on the choice of organic food products. Such antecedents are important when individuals feel a sense of obligation to buy ethically, while identifying themselves as ethical consumers. This study follows the line of enquiry first indicated by Shaw and associates in a study that has resulted in diverse articles – some of which analyzing the same data with different methods – in various journals (Shaw and Shiu 2002ab; Shaw, Shiu and Clarke 2000). In these, Shaw and associates examine the role of ethical obligations and self-identity in addition to those determinants predicted by the *Theory of planned behavior* (Ajzen 1991). The aim of these studies is to investigate the intention to buy organic food products in comparison with non-organic food products and, specifically, to understand the role of those ethical considerations which underlie the purchase of these products. Consistent with the extant research in this field (e.g., Michaelidou and Hassan 2008; Thøgersen 1999), these studies describe moral norms in terms of personal beliefs regarding what is right or wrong, and ethical self-identity as the degree to which one perceives oneself to be an ethical consumer and, as such, to be concerned about ethical issues (Sparks and Guthrie 1998). Results show, among other things, that moral norms and self-identity have a direct influence on the intention to buy organic food products; furthermore, they both have an indirect effect on purchase intention through attitude, confirming the findings of Shaw and opening the way to the verification of the role of attitude as a mediating variable in the effects of moral norms on the intention to buy, a prospect that reveals itself more than once in this book.

In Chapter 3, the role of other determinants of Ajzen's (1991) *Theory of planned behavior* is examined (including the dimensions of *ethical products' image*). Self-imposed ethical obligations, in particular, have been re-conceptualized by using the construct of *moral disengagement.* This term refers to those cognitive mechanisms which allow individuals peace of mind when their behavior infringes their own moral codes (Bandura 1999). Moral disengagement is a sort of hypocrisy towards oneself, a kind of Machiavellianism (cf. Rawwas, Patzer and Klassen 1995) directed towards oneself, by which an individual displays a kind of cool detachment that makes him or her less emotionally involved with others or with ethics, thus allowing him or her to save face – first with him or herself – in potentially embarrassing situations, when this contradicts his or her own moral standards.[4] Findings in Chapter 3 stress the importance both of ethical dimensions

4 This chapter constitutes, from a different point of view, an empirical test of the hypotheses advanced by the *Neutralization theory* (Chatzidakis, Hibbert and Smith 2006), according to which, consumers can mitigate the negative impact of their ethically-questionable activities and of the psychological tensions that arise when they behave in apparent contradiction to their expressed ethical concerns by the use of justifications that soften or eliminate the impact that norm-violating behavior might have upon self-concept and social relationships. *Neutralization theory* takes up the issues of moral disengagement and adapts them to ethical consumption.

in the purchasing intention of organic food products and of the means employed to collect information regarding the motivation behind ethical consumption. It proposes, at the same time, an intercultural comparison between results in Italy and France for latent traits of organic food product image – identified through the *Five-factor model* of personality (cf. Digman 1990) and integrated in the *Prospect method* (Caprara and Barbanelli 2000; Caprara, Barbaranelli and Guido 2000) – which have an impact on ethical determinants. Regarding the traits of product image, the study allows one to examine questions related to the ethical dimension of shopping, analyzing them not as an end-value which is alternative to the classical hedonic or utilitarian dimensions pursued in shopping (see Babin, Darden and Griffin 1994; Shaw and Shiu 2003), but as a subcategory of these two dimensions. In other words, from the results reported in this chapter, one can infer that the ethical objectives – or, on the contrary, the individualistic ones – of consumers are comprised, at the same time, according to the dominant profile of the sample, within the utilitarian, self-interested aims sought in the shopping or, rather, within the hedonic, more societal-centered aims that may also prevail. Managerial and theoretical implications are also discussed.

Chapter 4 tries to find an answer to the question: "How does culture affect ethical purchasing rationale?" Belk, Devinney and Eckhardt (2005) assert that ethical interpretations may differ in the diverse consumer markets around the world (even if, occasionally, differences in socioeconomic position in society and even level of education may have less effect on perceptions of consumption ethics than expected). "Because ethics are a part of culture, to study ethical choices without explicitly considering the cultural context is not realistic. Differing cultural reactions to consumption practices would be expected to occur, not only because moral values are socially and culturally constructed, but also because there are cultural differences in social roles, gender roles, institutional structures, welfare expectations, laws, and traditional rights, privileges, and obligations. In other words, culture filters our perceptions of what constitutes good or responsible consumption and what are perceived to be the consequences of violating these moral norms" (Belk, Devinney and Eckhardt 2005: 278-279). Consumers' worries over food safety issues (*food safety concerns*) and degree of identification in the ethical consumer image (*ethical self-identity*) were recognized as the main motives for buying organic food, by acting directly on intentions and, again, indirectly, by the mediation of attitudes. Since very few studies have attempted to single out the existence of cultural differences in the pattern of organic food consumption, a survey conducted in the United Kingdom (Michaelidou and Hassan 2008) was replicated in Italy in order to investigate how the aforementioned constructs affect consumers' purchase intention cross-culturally. The differences in the findings of the two studies were interpreted adopting Hofstede's (1983, 1984) intercultural approach by considering, in particular, the dimensions of *Uncertainty avoidance* and *Individualism*, which express, respectively, consumers' reluctance in the face of unknown situations (such as possible food hazards connected to non-organic food products) and unwillingness to interact with other people (such as carelessness of other people's needs apart from one's own).

The study in Chapter 5 compares the level of acceptance of genetically modified food products on the part of a specific age group of consumers living in two different European countries. By administering a questionnaire which was previously tested in Norway (Honkanen and Verplanken 2004) on an Italian student sample, the influence of cultural factors was appraised in the decision-making process that leads consumers to support or to reject genetically modified organisms (GMOs), i.e. new varieties of breeding plants and animals created by the transferring of genetic material between different species (cf. Suslow, Thomas and Bradford 2002). Schwartz's (1992) model was applied to establish how *personal values* affect subjects' attitudes and purchasing intention with regard to such products, whereas the cross-cultural differences between respondents' behavioral patterns were explained with reference to Hofstede's (1983) cultural dimensions. Results confirmed Shaw *et al.*'s (2005) findings according to which non-GMOs consumers are either opposed to, or attach little importance to, Schwartz's (1992) *Power* values, which reflect an individual's need for dominance and control over people and resources; whereas these goals were found in conflict with their strong *Universalism* values, such as equality, peace, harmony, social justice, tolerance, protection of the environment. Suggestions for segmenting the demand for genetically modified food products according to psychographic criteria were also proposed, always bearing in mind that one should not use as a starting point false premises. As Newholm and Shaw (2007) maintain, the uncritical study of "ethical consumption" presupposes an everyday "unethical" consumption, which is not always confirmed by the facts. In this case, those who use GM foodstuffs are not necessarily unaware of the associated risks, but simply consider these risks as sufficiently low as to not be a threat to their or their loved ones' health. In reality, many ordinary acts of consumption are normally subject to ethical control on the part of the moral standards of the buyer, even if less "sophisticated" consumers may be – and, it seems indeed that they are (Titus and Bradford 1996) – more reluctant to penalize those companies that make use of unethical business practices. In the final analysis, the existence of an "ethical consumer" as distinct from a "traditional consumer" seems to be more of a myth than a reality (cf. Carrigan and Attalla 2001), because the ethical profile of consumer choices can emerge both when rational and utilitarian objectives are pursued, and when hedonic and social aims are. Rather, a lack of ethical awareness may stem from a disregard for goals and be attributable to a lack of adequate information which hampers the formation of an individual's beliefs (Shaw and Clarke 1999) or to a insufficient degree of personalization in the information, which creates a gap between attitudes and behaviors or which puts little pressure on businesses to become "actively ethical" ("Is there such a thing as an ethical consumer?" 2003).

Close examination of the beliefs on which consumer's social and environmental behaviors and perceptions are based is the subject of Chapter 6. In it, convictions which distinguish expert from novice consumers are examined when choosing alternative energy sources and low-emission vehicles. In particular, this conclusive chapter presents two related studies which lay down the premises for more detailed studies into the antecedents of the intention to use *bio-fuels*, namely, any solid, liquid or gaseous fuel derived from biological materials (i.e., biomasses) including food products (such as corn, soybean and other crops), and *hydrogen* as a clean source of energy for cars. Pollution, mainly due to the emissions of an increasing number of vehicles, and the international energy issue, arising from the progressive depletion of the world's oil reserves, which causes an obvious rise in the price of traditional fuels, both encourage the development of new technologies based on alternative and renewable energy sources (e.g., Baykara 2005). The aim of this chapter is to explore and gain insights into principal *beliefs*, which is to say the personal convictions derived from potential buyers' and industry experts' own specific knowledge of the subject regarding the introduction of these new sources of energy. The open-question phase of the *Theory of planned behavior* (Ajzen 1991) is used as a theoretical model of reference: the results obtained constitute the point of departure for future research designed to test – on experimental bases – the model, thus allowing to identify priorities for future energy policies. From the perspective of the production of bio-fuels, results highlight the ethical problems that underlie the cultivation of crops for bio-fuels, which could be used to feed starving populations (above all in developing countries), set against the obtainable advantages, still on a social level, of economic growth linked to the development of new technologies and freeing nations from the dependence on petroleum-based products. From the perspective of the launch of new car engines which maintain their original design and use hydrogen propellants, results allow one to gauge potential customers' degree of knowledge and predisposition towards these new products, as their judgments and perceptions of dimensions of product image are relevant considerations affecting the launch of such vehicles onto the market.

1.3. Conclusions

One of the problems of modern society is that dominant levels of materialism and high ethical standards are inversely correlated constructs (Muncy and Eastman 1998): the importance consumers attach to their worldly possessions, the centrality that these have in their lives, often leads individuals to suppress urges to behave in line with that which seems to be ostensibly the common good. These selfish motives, all too human and understandable in the capitalist society in which we live, which permeate an individual's every action and induce, through the constant bombardment of advertising and diffusion through the media of preeminently individualistic models of behavior, make people reluctant to leave the materialistic herd (e.g., Codeluppi 2003) and encourage them to keep on making, out of habit, morally questionable choices.

The possibility to take into consideration the existence of ethically-compatible purchase alternatives is, therefore, determined by the social and environmental context in which such alternatives may occur and, in particular, by the *salience* of these marketing stimuli (Guido 2001) – that is to say, by their preeminence of ethically-compatible alternatives in the perception of potential buyers. The need to make some information "salient" to incentivize ethical choices emerges clearly from recent studies. According to Thøgersen (1999: 440), for instance, "it is a necessary condition for the spread of moral environmental reasoning to buying decisions that characteristics which connect the purchase to environmental problems become *salient* in the buying situation. This is most likely to happen if *other* characteristics of the purchase are not too highly involving and, hence, are unable to 'monopolize' the attention of the consumer."

From this perspective, a relevant role is played by businesses which view consumers as the real "final products" of the transformation processes started with manufacturing, because advertising and fruition of goods "change" consumers' expectations and desires, as well as their perceptions; thus, in a way, "transform" people (Guido 1998, 2005; Varaldo and Guido 1997). Businesses' communication and sales strategies, therefore, should not be devised to either deceptively manipulate consumers' cognitive schemata or to dominate their desires. Rather, they should be used as a method to guide customers towards realistic expectations about a company's products, teaching them how to secure the greatest advantages from their purchases to the benefit both of the individual and of society. In general, relationships between businesses and customers should be tuned according to the number and the nature of the interests of the subjects involved. Ethical questions should be raised to reconcile or to promote different (sometime opposite) goals, not to act to the detriment of consumer health or of the higher interests of people.

Gianluigi Guido, Piermario Tedeschi, M. Irene Prete,
Laura Franceschini and Carolina Buffa

2. The influence of moral norms and self-identity in the choice of organic food products[1]

2.1. Introduction

Changes that have occurred in Western consumers' lifestyles and habits over the past few years have excited a growing interest in healthier models of life behavior and respect for the environment (Chinnici, D'Amico and Pecorino 2002; Wier and Calverley 2002). This accelerated trend has contributed to the diffusion of *organic food products*, which are produced using biological methods which avoid toxic pesticides and genetic engineering, and, consequently, are seen by consumers as safer for health and for the environment (cf. Rimal, Moon and Balasubramanian 2005). In recent years, the organic food market has registered a significant expansion: in many countries, this growth in consumption has been superior to the average registered in the food sector and, for this reason, the organic-food trade has been often considered as one of the most promising markets (Hughner *et al.* 2007). According to the Organic Trade Association (2007), in 2006, sales of organic products in the United States increased by 20.9%, against 7.4% of total food sales. In the same year, the global organic market showed a growth equal to 13.6%, achieving 36.7 billion dollars (Datamonitor 2007).

Considering the potential growth of this market, the analysis of the variables influencing consumers' choice can offer academicians a

1 The Authors would like to thank Daniela Battista and Massimo Sardi for their support in data collection and codification.

deeper understanding of individuals' expectations and desires in the choice of food products; furthermore, it can provide companies and institutions with information for marketing purposes, in general, and, for identifying appropriate entry strategies into this market, in particular. Several studies, aimed at analyzing the more frequent motives for the purchasing of organic products, have shown that they imply considerations concerning, on the one hand, advantages for the environment and the well-being of animals, and, on the other, personal ethics and the idea that consumers have of their own selves (Arvola *et al.* 2007; Baker, Thompson and Engelken 2004; Grunert and Juhl 1995; Magnusson *et al.* 2003; Zanoli and Naspetti 2002). The consumption of these products seems to be the result of an ideology, linked to a particular set of moral values (Schifferstein and Oude Ophuis 1998), which has consequences for health care, environmental protection, and moral responsibility toward consumers.

The aim of the present study is to analyze the purchase behavior of organic food products and conventionally-produced alternatives (i.e., non-organic products), with the objective of identifying determinants affecting the intention to buy them, and, specifically, to verify the influence of moral and self-identity dimensions on their purchase intention. Ajzen's (1991) *Theory of planned behavior* (hereafter TPB) was used, according to which the *Intention* to buy organic or conventional food products is a decision influenced by: *Attitude*, i.e. favorable or unfavorable propensity towards this behavior; *Subjective norm*, i.e. pressure of important others in favor of or against it; and *Perceived behavioral control*, i.e. easiness or complexity of adopting it. As the TPB inadequately describes the moral dimension (Armitage and Conner 2001; Arvola *et al.* 2008; Sparks and Shepherd 2002) which can affect the *intention* to buy organic food products, two additional predictors were included in an extended version of the model: *Moral norms*, i.e. personal beliefs regarding what is right or wrong (Parker, Manstead and Stradling 1995); and *Self-identity*, i.e. relatively stable characteristics that individuals attribute to themselves (Sparks and Guhtrie 1998).

The present chapter is structured as follows: after defining organic food products, Section 2 reviews literature concerning motives and obstacles in purchasing them, by presenting the TPB as the standard theoretical framework of reference, as well as analyzing two further determinants of purchase intention (namely, *moral norms* and *self-identity*). Section 3 deals with research objectives, hypotheses, methodology, as well as results obtained from data analysis. Lastly, Section 4 provides a discussion of results, implications, limitations and directions for future research.

2.2. Organic food products, moral norms and self-identity

2.2.1. The nature of organic food products and organic agriculture

Organic food products are those products made by avoiding the use of chemical substances and by respecting the environment. They involve a farming method that, according to OTA (Organic Trade Association 2009: 1), "maintains and replenishes soil fertility without the use of toxic and persistent pesticides and fertilizers." Notwithstanding the fact that consumer literature shows a sufficient knowledge and awareness of organic products and provide detailed reviews which compare nutritional and safety related attributes of organic and conventional food products (Bourn and Prescott 2002; Woese *et al.* 1997), consumers on the market often show confusion and discrepancies in their understanding of organic food and agriculture (Hill and Lynchehaun 2002; Hutchins and Greenhalgh 1997; Wolf 2002). Some studies (e.g., Antonelli 1996) demonstrate that their definition on the part of consumers is often vague and inaccurate, also due to an incomplete knowledge of their nature. Actually, these products possess *credence qualities* – characteristics that cannot be detected or evaluated before their purchase or by ordinary use only (Andersen and Philipsen 1998; Darbi and Karni 1973; Giannakas 2002; Hansen 2001). To clarify this theme, in 2005, the International Federation of Organic Agriculture Movements (IFOAM) set up a "task force" designed to identify a concise and shared definition of organic agriculture, in order to provide a consumption standard. The process, concluded in March 2008, defined *organic agriculture* as "a production system that sustains the health of soils, ecosystems and people. It relies on ecological processes, biodiversity and cycles adapted to local conditions, rather than the use of inputs with adverse effects. Organic agriculture combines tradition, innovation and science to benefit the shared environment and promote fair relationships and a good quality of life for all involved" (IFOAM 2009: 1).

Organic agriculture is based on four principles stated by the General Assembly of IFOAM (2005: 1): (*i*) *health*, as organic agriculture "should sustain and enhance the health of soil, plant, animal, human and planet as one and indivisible"; (*ii*) *ecology*, as it "should be based on living ecological systems and cycles, work with them, emulate them and help sustain them"; (*iii*) *fairness*, as it "should build on relationships that ensure fairness with regard to the common environment and life opportunities"; and (*iv*) *care*, as it "should be managed in a precautionary and responsible manner to protect the health and well-being of current and future generations and the environment".

2.2.2. Motives for organic food purchase

Several studies have investigated the comparison between organic and conventionally produced food, from the viewpoint of both companies – in terms of production, pricing, and profitability differences – and consumers – in terms of nutritive, sensory, and safety characteristics (see Bonti-Ankomah and Yiridoe 2006, for a review). In particular, from the consumer perspective, there is disagreement over whether organic products are better than conventional products, and whether they possess different features or not, and therefore, what features differentiate them. As organic food are *credence products*, consumers may not distinguish between organic and conventionally produced food (Giannakas 2002), therefore, their willingness to pay a premium price for organic products may be very low. As sensory features alone may not suffice for their evaluation, quality signals, such as organic product label and brands, can contribute to the differentiation of such products, although consumers are often skeptical because of "mislabeling, and product misrepresentation, and partly because of non-uniform organic standards and certification" (Bonti-Ankomah and Yiridoe 2006: 38).

In consumer literature, several studies have been conducted in order to identify *personal motives* and *external factors* encouraging organic food purchase. As regards to *personal motives*, they can be classified according to *individual* and *selfless* reasons (e.g., Michaelidou and Hassan 2008).

Individual motives, such as taste, food safety and health care, have been found to be the most relevant factors determining organic-food purchase. According to McEachern and McClean (2002), the majority of people believe that *taste*, defined as the degree of genuine flavor, consistency and smell of organic products (Padel and Foster 2005), allows them to appreciate their value (Zanoli and Naspetti 2002). Secondly, people buy them because of *food safety*, which refers to the degree to which organic products are perceived as trustworthy, certified and natural (Zanoli and Naspetti 2002), as well as deprived of chemical residuals (Michaelidou and Hassan 2008; Williams and Hammitt 2001), and not genetically modified (Rimal, Moon and Balasubramanian 2005; Roselli, Seccia and Stasi 2006). Thirdly, they are purchased because of *health care* reasons, i.e. the degree to which the absence of pesticides and chemical substances in organic products contributes to avoid diseases (Padel and Foster 2005): as a matter of fact, the majority of consumers consider these products safer, healthier and more nourishing than conventional products (Lea and Worsley 2005; Williams and Hammitt 2001), nevertheless, to this day, there is no accepted evidence of this (Magkos, Arvaniti and Zampelas 2003, 2006; McKay 2007). Literature does not agree on which individual motive is predominant in the purchase of organic products: some studies highlight the importance of taste (McEachern and McClean 2002; Zanoli and Naspetti 2002), others food safety (Michaelidou and Hassan 2008; Rimal, Moon and Balasubramanian 2005), and some others health care (Chinnici, D'Amico and Pecorino 2002; Lockie *et al.* 2002; Magnusson *et al.* 2003; Makatouni 2002; Padel and Foster 2005; Zanoli and Naspetti 2002).

The other category of motives includes *selfless* motives: *environmental protection*, the degree to which organic agriculture has a low impact on the environment (Zanoli and Naspetti 2002), contributes to reduce pollution and uses non-renewable resources responsibly (McEachern and McClean 2002); *animal welfare*, the degree to which humane breeding methods (Makatouni 2002; Schröder and McEachern 2004) and the absence of drugs and hormones in food production contribute to improving animal well-being and health, as well as to respecting their rights (Honkanen, Verplanken and Olsen 2006; McEachern and McClean 2002; Padel and Foster 2005); *political motivations*, both the degree to which organic products contribute to promoting local economy and fair-trade (Padel and Foster 2005), and their originating in politically-acceptable countries, where human rights are not violated (Honkanen, Verplanken and Olsen 2006); *moral responsibility towards users*, the degree to which organic food purchase is influenced by moral considerations and by the responsibility purchasers of organic products feel that they have towards users, as happens for products purchased for relatives (Makatouni 2002; Raats, Shepherd and Sparks 1995). For example, according to Finch (2005), such events as pregnancy might lead to the purchase of organic products. There is disagreement in literature also on predominant selfless motives: whereas some studies lean toward environmental motivations and animal health (Honkanen, Verplanken and Olsen 2006; Lockie *et al.* 2002), others highlight moral responsibility toward family members' health and well-being (Makatouni 2002; Raats, Shepherd and Sparks 1995).

Both these categories of motives tend to positively influence the individual's intention to purchase organic products. Nevertheless, the percentage of people really buying organic products turns out to be much lower than that of individuals who declare that they purchase them; this might be due to *external factors* which thwart purchase of organic products and, simultaneously, encourage the preference of conventionally produced alternatives. Consumer *awareness and knowledge* about organic products are crucial factors in the individual's purchase decision. Whereas habitual customers or well-informed consumers may be aware of the differences between the two alternative products (Giannakas 2002), the lack of knowledge can induce consumers to prefer the cheaper alternative. Indeed, *price* has also a relevant importance; it refers to the degree to which the premium price applied to organic products hinders their consumption (Lea and Worsley 2005; Padel and Foster 2005; Zanoli and Naspetti 2002). This obstacle does not represent an absolute barrier, as consumers would be disposed to pay a higher price for specific advantages of organic products, according to some studies in North America (Buzby and Skees 1994; Goldman and Clancy 1991; Harris 1997; Jolly 1991), Western Europe (Canavari *et al.* 2002; Chinnici, D'Amico and Pecorino 2002; O'Donovan and McCarthy 2002; Padel and Foster 2005; Soler, Gil and Sanchez 2002) and the rest of the world (Aguirre 2001; Wang, Halbrendt and Webb 1997). Another external factor is their *difficult availability*, referring to the degree to which organic products are more difficult to find in retailing stores (Lea and Worsley 2005; Zanoli and Naspetti 2002). Indeed, whereas in such countries as Norway, Denmark or the United Kingdom organic products are largely present in large-scale retail trade, in some other nations, such as Italy, France and Germany, they are mainly distributed in less diffused retailing channels, such as specialized stores (De Ruvo 2007).

2.2.3. The identification of determinants influencing organic food purchase: The Theory of planned behavior and its extension

Consumers often associate concerns of ethics and sense of responsibility for the well-being of their relatives and the safeguard of nature: indeed, motives shaping the purchase of organic food products are often related to food safety and health care reasons, environmental protection, and moral responsibility toward users. So, it could be interesting to deepen the understanding of moral evaluations related to the purchase of organic products, and, in particular, to determine which peculiar motives can affect it.

The theoretical framework used in this study to evaluate the impact of moral dimensions on organic food purchase intention is Ajzen's (1991) *Theory of planned behavior* (TPB), one of the most consolidated models in social psychology. The TPB represents an extension of the *Theory of reasoned action* (TRA, Ajzen and Fishbein 1980; Fishbein and Ajzen 1975) and was formulated in order to overcome the limit of the TRA, applicable only to behaviors completely depending on the will of individuals. The basic assumption of both these theories is that the *intention* of adopting a specific behavior (e.g., the intention to buy organic food products) is the best predictor of that behavior. In the original formulation of the TPB (Ajzen 1991), *intention* is a function of three principal determinants: *attitude toward behavior*, i.e. the level of an individual's favorable or unfavorable propensity for the considered behavior; *subjective norm*, i.e. the degree of social pressure that important others exert on individuals, leading them to adopt or not a specific behavior; and *perceived behavioral control*, i.e. the individual's perception of the level of easiness or difficulty of accomplishing that specific behavior. According to this model, each of these determinants of the behavioral *intention* is based on *salient beliefs*, that is, all the relevant mental associations between that behavior and its perceived characteristics (for a criticism of this definition of "salience", see Guido 2001). In line with the so-called "expectancy-value" models (cf. Eagly and Chaiken 1993, for a review), *intention*'s determinants are obtained by multiplying the subjective probability of the occurrence of some consequences related to the behavior (the expectancy) by the corresponding evaluation (the value) of such a behavior (see Figure 2.1, *below*). Therefore, *attitude* results from the so-called *behavioral beliefs*, that is, the sum of products of the subjective probability that the target behavior will assure specific advantages (or disadvantages) and the relative importance assigned by the perceiver to each of these consequences. *Subjective norm* results from *normative beliefs*, that is, the sum of products of the subjective probability that the behavior will be approved (or disapproved) by significant others and the corresponding subjective motivation to comply with them. Likewise, *perceived behavioral control* results from *control beliefs*, that is, the sum of products of the subjective probability of certain events that could facilitate (or hinder) the behavior and the corresponding evaluations of the importance of such events in influencing the behavioral intention.

According to Ajzen (1991), the TPB is open to the inclusion of additional predictors when they explain a significant proportion of the variance of behavioral intention. Several studies have been carried out in order to verify the validity of the extension of the TPB to other additional determinants – whose importance varies according to the

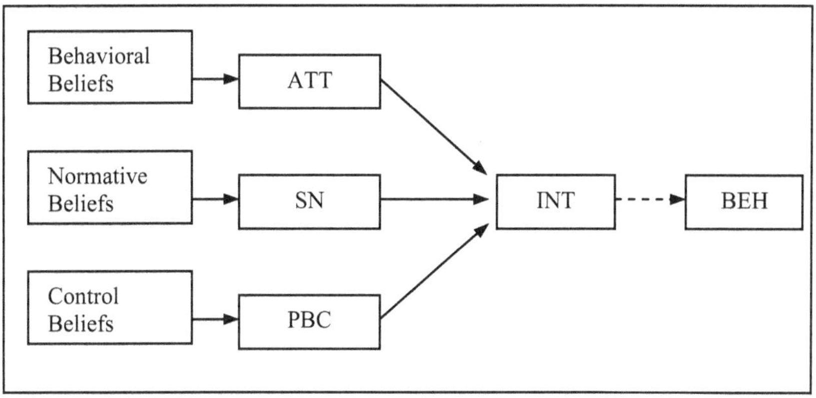

Note: ATT = Attitude; SN = Subjective norm; PBC = Perceived behavioral control; INT = Intention to perform a specific behavior; BEH = Behavior (e.g., the intention to buy organic food products).

Figure 2.1. The influence model in the *Theory of planned behavior*.

behavior considered (see, for example, Conner and Armitage 1998; Langdridge, Sheeran and Connolly 2007).[2]

In the present study, *moral norms* and *self-identity* were added in the TPB as useful variables in order to examine the role of consumers' moral evaluations and self-identities in organic products purchase (e.g., Arvola *et al.* 2008; Raats, Sheperd and Sparks 1995; Shaw *et al.* 2005; Sparks and Sheperd 2002). Considering the ethical nature of some of the most frequent motivations in organic food purchase, *moral norms*, i.e. personal beliefs about right and wrong (Parker, Manstead and Stradling 1995), and *self-identity*, i.e. relatively stable characteristics that individuals attribute to themselves could have a significant influence. As *moral norms* are individual value judgments con-

2 Normally, the TPB is extended to *Past behavior*, considered a good predictor of future behavior, if internal and external factors determining it remain stable (Jaccard and Blanton 2005). *Past behavior*, considered a good predictor for habitual consumption, such as in the purchasing of food products, has been found to be a significant determinant in Thøgersen's study (2002), which highlights that consumers having a previous experience with organic food products usually continue to buy them.

nected to a specific behavior and to its outcomes (Manstead 1999), they can play an important role, not only in the case of ethical behaviors (Conner and Armitage 1998; Manstead 1999), such as respect for others and individual accountability, but also in the case of those behaviors implying less involvement, such as food product purchase (Goldsmith, Frieden and Henderson 1997; Homer and Kahle 1988). Indeed, *moral norms* have been one of the variables frequently used to extend the TPB in the context of behaviors with an ethical dimension, such as eating genetically-modified food (Spence and Townsend 2006), breaking the speed limits (Conner *et al.* 2007; Conner, Smith and McMillan 2003), behaving fairly in one's profession (Buchan 2005), hurting oneself deliberately (O'Connor and Armitage 2003), taking drugs (McMillan and Conner 2003). Also *self-identity* has been used to extend the TPB (e.g. Biddle, Bank and Slavings 1987), especially to explain behaviors in which the consumer's degree of identification with specific behavioral models is relevant, such as going on a low animal-fat diet (Sparks and Guthrie 1998), eating fruit and nuts (Dennison and Shepherd 1995), buying fair-trade products (Shaw, Shiu and Clarke 2000), donating blood (Charng, Piliavin and Callero 1988) and voting (Granberg and Holmberg 1990).

2.2.3.1. The Theory of planned behavior and the purchase of organic food products

In consumer literature, several studies have used the TPB in order to assess the importance of the determinants of intention to buy organic food products (see Table 2.1, *below*). In particular, some articles have confirmed that positive *attitude* toward organic products affects *intention* more than other variables (Michaelidou and Hassan 2008; Sparks and Shepherd 1992; Thøgersen 2002). Although the link with the behavior has been found to be weak, due to the obstacles hindering purchase, such as price and availability of these products (Chen 2007; Lofordos and Dennis 2008; Sparks and Shepherd 1992), some other studies (Arvola *et al.* 2008; Tarkiainen and Sundqvist 2005) have obtained opposite results, showing that *perceived behavioral control* –

Author (Year)	*Countries*	*Method and Sampling*	*N*	*Variables*	*Products*
Arvola *et al.* (2007)	Finland, Italy, and UK	Self-administered questionnaire	270 (F), 202 (I), 200 (UK)	MA*, ATT*, SN*, PBC	Organic apples and ready-to-cook pizza
Chen (2007)	Taiwan	Self-administered questionnaire by using stratified sampling	470	ATT*, SN*, PBC*, FRPT* moderating effect on ATT and SN	Organic food
Dean *et al.* (2006)	Finland, Italy, and UK	Convenience sample randomly assigned to one of the three different tasks	91, 90, 110	Comparison of methods of eliciting emotions and beliefs	Fresh organic food
Honkanen, Verplanken and Olsen (2006	Norway	Questionnaire sent by mail to a sample selected by phone	1283	ATT*, EV* (indirect influence through ATT)	Organic products
Lodorfos and Dennis (2008)	UK	Questionnaire administered to a random sample	144	ATT*, SN*, PBC*	Organic products

Table 2.1. Synoptic table of main studies on the TPB and organic products (continues).

Author (Year)	*Countries*	*Method and Sampling*	*N*	*Variables*	*Products*
Michaelidou and Hassan (2008)	Scotland	Questionnaire administered to a random sample	222	ATT*, SI* (direct and indirect influence through ATT)	Organic products
Sparks and Shepherd (1992)	England	Questionnaire sent by mail to a random sample	261	ATT*, SN*, PBC*, SI*	Organic vegetables
Tarkiainen and Sundqvist (2005)	Finland	Questionnaire administered to a random sample at a hypermarket	200	ATT*, SN* (indirect influence through ATT), PBC	Organic flour and bread
Thøgersen (2002)	Denmark	Questionnaire administered to wine consumers randomly selected in malls	214	ATT*, SN, MN*, PB*	Organic red wine
Thøgersen and Ölander (2006)	Denmark	Panel of a random sample interviewed three times	1520	ATT*, SN*, MN*	Organic products

Note: * = Significant influence on intention; ATT: Attitude; EV: Ethical values; FRPT: Food-related personality traits; MA: Moral attitude; MN: Moral norms; PBC: Perceived behavioral control; PB: Past behavior; SI: Self-identity; SN: Subjective norm.

Table 2.1. Synoptic table of main studies on the TPB and organic products.

based on the perception of availability and price of organic food – does not significantly influence *intention*. Also *subjective norm* can exert, by the social pressure of "important others", a direct – though weak – influence on intention to buy organic products (Sparks and Shepherd 1992). Moreover, Tarkiainen and Sundqvist (2005) have proved its indirect influence through *attitude*. Yet, the opposite results were presented by Thøgersen (2002), who examined the purchase of organic red wine and found that *subjective norm* is not a significant predictor of *intention*. In conclusion, there is no agreement in literature on the existence of a predominant effect, among the three conventional predictors included in the TPB.

Other studies have extended the TPB by considering additional determinants, in order to improve the predictive capacity of the model in the context of organic food purchase, but the results they have obtained remain conflicting. As regards *moral norms*, their inclusion in the TPB, which seems necessary for the purchase of organic food (Arvola *et al.* 2008; Conner and Armitage 1998; Manstead 1999), has originated two streams of research: one examines the influence that *moral norms* directly exert on *intention*; the other stream of research studies the indirect influence *moral norms* exert on *intention*, considering *moral norms* as antecedents of *attitude*. As far as the former stream is concerned, in some studies the inclusion of *moral norms* has improved the capacity of the TPB of predicting purchase behavior toward organic food (Thøgersen 2002; Thøgersen and Ölander 2006); as regards the latter stream of research, literature has focused on the analysis of personal values influencing *attitude* towards organic food purchase. In particular, according to Dreezens *et al.* (2005), Grunert and Juhl (1995), the value of *Universalism* – which can be identified through the principal motivational objectives, i.e. understanding for the environmental and people protection (Schwartz 1992) – has a positive influence on *attitude* towards organic products. However, Honkanen, Verplanken and Olsen (2006) demonstrated that ethical motivations linked to environmental and religious considerations influence *attitude* towards organic products. At the same time, *self-identity* has been usefully included in the TPB in the case of organic products, as the influence of *self-identity* on *intention* has been shown to increase when the frequency of a behavior is high. Thus, some studies on the extended TPB have demonstrated that *self-identity* affects *intention* to purchase organic products both directly (Michaelidou and Hassan 2008; Sparks and Shepherd 1992) and indirectly, through *attitude* (Michaelidou and Hassan 2008).

2.3. Aim and objectives

The purpose of the present study is to analyze the purchase behavior of organic food consumers, in order to assess if *moral norms* and *self-identity* play a predominant role among the other determinants of *intention* to buy and to identify which moral considerations have a stronger influence on purchase intention. Specifically, we hypothesize that *moral norms* – related to environmental protection, naturalness of food, and concern for the health of oneself and of one's loved ones – as well as consumers' *self-identity* – related to concerns with one's figure and health, awareness of the benefits of organic products, and consequences of one's purchasing for the development of the territory where one lives – influence consumers' *intention* to purchase organic food products.

2.3.1. Research hypotheses

The two constructs considered, *moral norms* and *self-identity*, have been included in the TPB (Ajzen 1991) as further determinants of *intention* to purchase organic products, together with *attitude*, *subjective norm*, and *perceived behavioral control*, as other authors did in the past (e.g., Chen 2007; Honkanen, Verplanken and Olsen 2006; Lofordos and Dennis 2008; Michaelidou and Hassan 2008; Sparks and Shepherd 1992; Thøgersen 2002; Thøgersen and Ölander 2006).

Specifically, we hypothesize that moral considerations and consumers' *self-identity* influence *intention* to purchase organic food products:

H1: Moral norms and self-identity will directly affect intention to purchase organic food products.

Furthermore, it has been additionally intended to verify whether *moral norms* and *self-identity*, as antecedents of *attitude*, have an indirect influence on *intention*. Individuals' moral norms and their *self-identity* may act, in fact, as antecedents not merely of purchase intention of organic food products, but also of *attitude*. Indeed, some studies have stressed the importance of understanding the role of *self-identity* and *moral norms* in determining *attitude* (Shaw and Shiu 2002ab; Shaw, Shiu and Clarke 2000; Sparks and Guthrie 1998; Sparks and Shepherd 1992). Therefore, it is hypothesized that:

H2: Moral norms and self-identity will indirectly affect intention in purchasing organic food products through attitude.

Using the analysis of non-organic food products as a control tool, it has been also hypothesized that moral considerations and *self-identity* do not directly or indirectly motivate the purchase of conventional products:

H3: Moral norms and self-identity will not, directly or indirectly, affect intention to purchase non-organic food products.

2.4. Methodology and results

This study is divided into two steps: a *pilot study*, which focuses on the identification of the salient beliefs that are at the basis of *attitude*, *subjective norm*, *perceived behavioral control*, *moral norms*, and *self-identity*; and a *main study*, aiming to evaluate the impact (if any) of the above-mentioned determinants on the *intention* to buy organic and non-organic food products.

2.4.1. The pilot study

The pilot study was conducted in Rome (Italy) using a sample of 160 subjects, divided into two sub-samples of 80 subjects. Each sub-sample comprised 40 men and 40 women, half users and half non-users of organic food products. Respondents were gathered into five age-groups (15-29, 30-41, 42-53, 54-65, 66+ years old), selected according to the direct interception method, i.e. the random selection of shoppers in supermarkets and specialized stores.

The objective of the pilot study was to identify the salient beliefs needed for the development of the main questionnaire for each category of products considered. To this end, two different open-ended questionnaires were administered, following Ajzen and Fishbein (1980; Fishbein and Ajzen 1975) suggestions for this type of research: one questionnaire concerning organic food products was administered to one half of the sample, and another questionnaire concerning non-organic food products was administered to the other half. Both questionnaires were aimed at explicating: *behavioral beliefs*, on which *attitude* is based, obtained by asking subjects to list advantages and disadvantages related to the purchase of organic and non-organic products; *normative beliefs*, on which *subjective norm* is based, by asking subjects to list those people or those groups of people that might approve or disapprove of their purchase of these products; *control beliefs*, on which *perceived behavioral control* is based, by asking subjects to make a list of facts or circumstances that could facilitate or hinder their *intention* to buy organic or non-organic products. The results of the pilot study, obtained by the content analysis carried out on protocols of answers and considering the most frequent ones, were used to develop two main questionnaires, one concerning organic food products and the other regarding non-organic ones.

2.4.2. The main study

The main study was conducted in Rome, Italy, using a random sample of 320 consumers – divided into two sub-samples of 160 subjects, grouped into the following age groups: 16-27, 28-39, 40-51, 52-59, and 60-75+ years old (see Table 2.2, for details).

Two different closed-ended questionnaires were administered: one concerning organic food products was administered to half of the sample, and another regarding non-organic food products was administered to the other half. Both questionnaires were composed of two parts: the first part employed a 7-point Likert scale concerning behavioral, normative, and control beliefs, to assess the constructs of the TPB (i.e., *attitude*, *subjective norm*, and *perceived behavioral control*); *moral norms* and *self-identity* were measured with two scales taken from, respectively, Armitage and Conner (1999) and Raats, Sheperd and Sparks (1995); whereas purchase *intention* was assessed by two items, on a 7-point Likert scale, measuring, respectively, the strength of the intention to buy organic food in the two weeks following the test and the subjective probability of effectively engaging in such a behavior over that specific period of time. The second part of the questionnaire collected socio-demographic data (i.e., gender, age, education, marital status, family size, favorite retail store).

Typology of product	*N*	*Gender*	*Age class*
Organic products	160	M = 64 F = 96	28 subjects 16-27 y.o. 55 subjects 28-39 y.o. 26 subjects 40-51 y.o. 26 subjects 52-59 y.o. 25 subjects 60-75+ y.o.
Non-organic products	160	M = 66 F = 94	39 subjects 16-27 y.o. 33 subjects 28-39 y.o. 26 subjects 40-51 y.o. 27 subjects 52-59 y.o. 35 subjects 60-75+ y.o.

Table 2.2. Details of the sample in the main study.

2.4.3. Results

In order to verify if the determinants of the extended TPB can be valid predictors of *intention* to buy organic products, a multiple linear regression analysis was used, by considering *intention* as the dependent variable, and *attitude, subjective norm, perceived behavioral control, moral norm*, and *self-identity* as independent variables. A hierarchical regression shows that the extended TPB model fits real data (R^2 = .610) and, furthermore, that the inclusion of *moral norms* and *self-identity* significantly improves the model. The determinants that significantly influence *intention* to buy organic products are *moral norms*, which have the highest beta value ($\beta = .343$, $p < .001$); then, *subjective norm* ($\beta = .250$, $p < .001$), *self-identity* ($\beta = .219$, $p < .001$) and *attitude* ($\beta = .136$, $p < .05$) (see Table 2.3). Thus, Hypothesis H1, according to which *moral norms* and *self-identity* have a direct influence on *intention* to buy organic food products, was confirmed.

Model	*Variables*	*B*	*Beta*	*t*	*p*
TPB	Constant	25.611		14.214	.000
	ATT	*.60*	*.356*	*5.057*	*.000*
	SN	*.72*	*.431*	*6.051*	*.000*
	PCB	-7.4E-005	-.001	-.007	.994
Extended TPB	Constant	20.394		12.427	
	ATT	*.023*	*.136*	*2.099*	*.038*
	SN	*.042*	*.250*	*3.927*	*.000*
	PBC	.008	.056	.936	.351
	MN	*.057*	*.343*	*4.304*	*.000*
	SI	*.025*	*.219*	*2.846*	*.005*

Note: Dependent Variable: Intention; N = 160; TPB model: R = .658; R^2 = .433; Adjusted R^2 = .442; Extended TPB model: R = .781; R^2 = .610; Adjusted R^2 = .596; ATT = Attitude; SN = Subjective norm; PBC = Perceived behavioral control; MN = Moral norms; SI = Self-identity.

Table 2.3. Comparison between the TPB and the extended version of the TPB for organic food products.

From the mean analysis of beliefs, which are the basis of the variables significantly influencing *intention* (Tables 2.4–2.6 and 2.7, *below*), it is possible to infer that individuals buy organic food products principally because they are aware of their naturalness (M = 33.906, SD = 10.700) and because they intend to protect their family members'

Relevant beliefs behind moral norms	M	SD
I would buy organic products to eat natural food	33.906	10.700
I would buy organic products to protect my relatives' health	33.162	11.193
I would buy organic products to protect my own health	32.906	12.321
I would buy organic products to protect the environment	32.444	11.754

Table 2.4. Salient beliefs affecting *moral norms* for organic food products.

Normative beliefs	M	SD
My partner (wife/husband, fiancée/fiancé) would approve	27.475	14.979
My relatives would approve	25.731	14.297
My friends would approve	24.275	14.093

Table 2.5. Normative beliefs affecting *subjective norm* for organic food products.

Relevant beliefs behind self-identity	M	SD
I think of myself as someone who enjoys the pleasures of good eating	32.444	11.431
I think of myself as someone who eats healthy and is concerned about his/her health	32.094	12.434
I am aware of the benefits of organic products	29.706	11.100
I am patient and careful in choosing food	29.706	11.802
I am concerned about my figure	28.219	13.079
I am a person that develops my local area through purchasing	27.294	12.562

Table 2.6. Salient beliefs affecting *self-identity* for organic food products.

Behavioral beliefs	M	SD
I would buy risk free and healthy products	29.220	12.372
I would buy natural, chemical and additives free products	29.113	11.516
I would enjoy healthy and balanced food	29.040	10.798
I should afford high prices when buying organic food products	24.566	11.302
I should face low variety/assortment for available organic food products	22.393	11.852
I should face difficulties in finding specialized stores of organic food products	20.980	10.556

Table 2.7. Behavioral beliefs affecting *attitude* for organic products.

health (M = 33.162, SD = 11.193). The social pressure they felt principally derives from their partners (M = 27.475, SD = 14.979) and relatives (M = 25.731, SD = 14.297). Organic food consumers think of themselves as individuals who enjoy the pleasures of good eating (M = 32.444, SD = 11.431), and are concerned about their own health (M = 32.094, SD = 12.434). So, they would mainly buy risk free and healthy products (M = 29.220, SD = 12.372).

In order to test the second hypothesis – i.e. to determine if *moral norms* and *self-identity* indirectly affect *intention* to buy organic food products – a linear regression analysis was conducted, by considering *attitude* as the dependent variable and *moral norms* and *self-identity* as independent ones. Results, reported in Table 2.8, lead one to accept Hypothesis H2, according to which *moral norms* ($\beta = .338$, $p < .001$)

Variables	*B*	*Beta*	*t*	*p*
Constant	84.140		8.257	.000
MN	*.338*	*.342*	*3.356*	*.000*
SI	*.144*	*.209*	*2.049*	*.042*

Note: Dependent variable: Attitude; $N = 160$; $R = .514$; $R^2 = .264$; Adjusted $R^2 = .254$; MN = Moral norms; SI = Self-identity.

Table 2.8. Results of linear regression analysis of *attitude* on *moral norms* and *self-identity* for organic food products.

and *self-identity* (β = .144, $p < .05$) can be considered relevant antecedents of *attitude* in the context of organic food purchasing.

In order to verify the third research hypothesis, the same analysis was carried out for non-organic products as well. Such a regression, aimed at identifying the determinants of *intention*, showed that *moral norms* and *self-identity* do not affect *intention* to buy non-organic products ($p > .05$) in both cases, and that the variables that positively influence *intention* are only *perceived behavioral control* (β = .613, $p < .001$) and *subjective norm* (β = .145, $p < .05$). The model fits adequately to real data (R^2 = .522), as the total variance explained is over 50% (see Table 2.9). Thus Hypothesis H3 can be accepted.

Indeed, the study of relevant beliefs concerning *perceived behavioral control* and *subjective norm* (see Tables 2.10-2.11, *below*)

Variables	*B*	*Beta*	*t*	*p*
Constant	11.602		5.304	.000
ATT	.010	.048	.783	.435
SN	*.033*	*.145*	*2.450*	*.015*
PBC	*.128*	*.613*	*9.610*	*.000*
MN	-.009	-.036	-.518	.605
SI	.021	.122	1.713	.089

Note: Dependent Variable: Intention; N = 160; R = .723; R^2 = .522; Adjusted R^2 = .507; ATT = Attitude; SN = Subjective norm; PBC = Perceived behavioral control.

Table 2.9. Results of linear regression analysis for non-organic food products.

Control Beliefs	*M*	*SD*
Low price and promotional sales	32.469	12.829
Higher variety/assortment of food	32.387	12.018
The presence of a store or supermarket near home	30.094	13.224
Short time on one's hands	19.162	12.405
Watching particularly attractive advertisements	17.137	11.847
Health problems caused by consumption of non-organic food	16.706	12.985

Table 2.10. Control beliefs affecting *perceived behavioral control* for non-organic food products.

Normative Beliefs	*M*	*SD*
My partner (wife/husband, fiancée/fiancé) would approve	26.887	14.225
My relatives would approve	24.256	13.856
My friends would approve	20.306	13.452

Table 2.11. Normative beliefs affecting *subjective norm* for non-organic food products.

showed that consumers buy non-organic products mainly for their low prices and promotional sales (M = 32.469, SD = 12.829), as well as for their variety (M = 32.387, SD = 12.018). Moreover, even in the case of conventionally produced products, partners (M = 26.887, SD = 14.225) and relatives (M = 24.256, SD = 13.856) exert a strong and significant pressure on the decision maker.

2.5. Effects for consumer psychology and marketing strategies

2.5.1. Discussion

Results confirmed the three hypotheses of the study. Specifically, the purchase of organic food products is affected by *moral norms* and *self-identity*, as well as by consumers' *attitude* and social pressure of important others. *Moral norms* and *self-identity* were found to be the variables having both a direct effect on *intention* to buy organic food products, and an indirect influence through *attitude*. This study also reveals that individuals' *moral norms* and *self-identity* do not play a significant role in affecting *intention* to purchase conventionally produced products, which is, differently, affected by *subjective norm* and *perceived behavioral control*.

The consumption of organic products does not seem to be linked only to the satisfaction of a functional need, but it has also an ethical dimension (Colacchio, Mastrangelo and Zarantonello 2008), as it implies a series of considerations related to the effects of organic food on health, sensorial pleasures, and environmental protection. Purchasing these products – considered healthier and more nourishing than conventional ones (Lea and Worsley 2005; Williams and Hammitt 2001) and made by using farming methods with a low environmental impact (Zanoli and Naspetti 2002) – allows consumers to conform to their moral norms and behave consistently with their self-identities. Unlike previous research – which has also analyzed the direct influence of *moral norms* (Honkanen, Verplanken and Olsen 2006; Thøgersen 2002; Thøgersen and Ölander 2006) and *self-identity* (Michaelidou and Hassan 2008; Sparks and Shepherd 1992) on *intention* to buy organic products – the present study highlights that these moral considerations influence the purchase behavior not merely through *intention* but also through *attitude* (cf. Honkanen, Verplanken and Olsen 2006). Individual *moral norms* and *self-identity* allow consumers to take into account the consequences coming from organic-food purchase and, therefore, contribute to shape their favorable or unfavorable propensity to buy these products. Furthermore, in the purchase of organic food products, consumers are affected not only by their moral dimensions – personal and internal(ized) ethical principles, driven by expected self-administered recompenses or punishments – but also by *subjective norm* – individual perception of external pressure exerted by important others on what is right or wrong (cf. Ajzen 1991; Arvola *et al.* 2008). This means that the social context where people live – especially partners, relatives, and friends – significantly influence their ethical choices.

As regards non-organic food, *moral norms* and *self-identity* are not found to be significant predictors of *intention* to buy: this finding suggests that consumers purchase conventional products, normally available in supermarkets and stores, not for ethical reasons related to altruistic or self-centered motivation. Moreover, buyers' choices are generally determined by such aspects as low price, availability and better assortment: as a matter of fact, *perceived behavioral control* is found to be the best predictor of *intention* to buy non-organic products. On the contrary, *perceived behavioral control* appears to be not significant for organic food, in contrast to findings from some previous studies (Lea and Worsley 2005; Padel and Foster 2005; Zanoli and Naspetti 2002), according to which high price, low availability and scarce assortment of organic products are negatively correlated with the intention to buy them. Two reasons can be adduced to justify these results: firstly, this study involves habitual organic food consumers, recruited in specialized stores – this suggests that they do not perceive high price and availability as obstacles to their purchase, since they have sufficient financial resources and time to go shopping to these stores; secondly, this study has the objective of measuring the impact of the determinants on *intention*, rather than the relationship between intention and effective purchase behavior for organic products, on which price and availability could probably have a stronger influence.

Finally, the analysis of relevant beliefs, on which *moral norms* and *self-identity* are based, contributes to clarify the motives underlying organic food purchase. This result shows that organic food consumers mainly consider such aspects as: (*i*) naturalness of food, both their own and relatives' health – thus confirming other findings (Chinnici, D'Amico and Pecorino 2002; Lockie *et al.* 2002; Magnusson *et al.* 2003; Makatouni 2002; Padel and Foster 2005; Zanoli and Naspetti 2002); and (*ii*) the perception of themselves as individuals that enjoy the pleasure of good eating, that eat healthily and health conscious (cf. Michaelidou and Hassan 2008; Sparks and Shepherd 1992). Unlike some studies establishing environmental protection as the principal motive for the purchase of organic products (Honkanen, Verplanken and Olsen 2006; Lockie *et al.* 2002), this study indicates that this motive is less important.

2.5.2. Managerial implications

The market for organic foods, though growing ever more rapidly, still constitutes a niche in the food sector (Carboni 2002). In this perspective, it is essential to understand organic food consumer behavior in order to provide companies with useful information. In particular, three typologies of possible strategies aimed at increasing the demand for organic food products can be identified. Firstly, it is possible to expand sales through a *penetration strategy* directed at actual consumers who, according to Antonelli (1996), can be classified as: "fundamentalists", characterized by a regular purchase behavior and a very favorable judgment of organic products; and "skepticals", characterized by a sporadic purchase behavior due to a low level of belief in the qualitative superiority of these products compared with conventional ones. To this end, companies should undertake marketing actions targeting each category of consumers, for example: promotions related to quantity, in order to increase the purchase frequency of "fundamentalist" consumers – sales of larger packs, competitions or buy three-pay for two offers – and specific communication campaigns designed to inform "skeptical consumers" of the benefits of organic products – distributing brochures at point-of-sales and intensifying public relations that contribute to the publication of articles in the specialized press, since these generally have more credibility than advertising. Secondly, companies operating in the organic food sector should focus on *market development* in order to reach another customer segment, the so-called "explorers" (Antonelli 1996), that is prospective consumers who show concern for health and the environment, but do not buy organic products yet. In this case, two possible strategies could be the creation of "organic food corners" in supermarkets – i.e. an area where to taste products free and find brochures – and of promotions, aimed at encouraging trials, such as discount tickets and free samples. Thirdly, in order to increase the demand for organic products it could be useful to carry out *product development* strategies, by extending the range of the products offered and increasing the purchases of actual customers through cross-selling, for example selling packages of combined products, such as organic milk and organic biscuits.

Results from the present study reveal that consumers' moral choices towards organic food are influenced by considerations related to naturalness of products, health, and self-perception; whereas price, availability, and assortment of these products do not affect significantly their purchase intention. So, in reference to product development strategies, it seems necessary to improve organic food products' distinctive characteristics, such as healthiness and naturalness, which consumers consider relevant. In order to reach a critical mass of customers it is not enough to target "green consumers" who buy organic products in order to protect the environment and animals (McEachern and McClean 2002), as this study has revealed this motivation is of secondary importance.

As quality is not an intrinsic characteristic of products – but a subjective construct depending on every single consumer's perceptions, desires and expectations (Guido 2005) – *communication*, which represents the principal tool for influencing individuals' perceptions, plays a critical role here. In particular, it is possible to identify some essential directions that companies could move in when planning their communication strategies. On the one hand, advertising messages should emphasize the promise of healthiness of organic products, without neglecting such aspects as their natural and authentic taste, in order to persuade consumers that organic products are functional with respect to the expression of their personal identities, since they are people who do not only eat healthy and are concerned about their health, but also enjoy the pleasure of good eating. In this regard, it has been demonstrated that communicating the positive effects deriving from organic products, rather than the risks posed by conventional products, is more efficacious in influencing intention to buy (Gifford and Bernard 2006). On the other hand, it would be appropriate to convey correct information on the fundamental principles of organic agriculture and how this alternative farming method makes it possible to obtain healthy and natural products. More accurate information about this sector seems to be necessary not only to extend the market to other consumer segments, but also to increase actual consumers' and especially "skepticals'" product loyalty. Therefore, it would be worthwhile improving the aspects enabling the identification of organic products, such as labeling and packaging, by making them comprehensible and attractive. Finally, as regards the media to be chosen for advertising, companies should prefer specialized magazines. As a matter of fact, according to Antonelli (1996), above all in Italy, most of consumers acquire information on organic products from specialized periodicals; moreover, exchanging ideas with friends and colleagues is a frequent source of information, whereas the mass media – such as newspapers and television – are less important.

2.5.3. Limitations and future research

The present study demonstrates a good predictive capacity of the TPB, extended to *moral norms* and *self-identity*, in the context of purchase behavior for organic products; nevertheless, like any survey, it has some limitations. The fact that questionnaires were administered in the presence of the interviewer may have led respondents – being subject to social desirability – to give socially-acceptable answers rather than answers based on their real opinions. Moreover, among the relevant beliefs considered to measure *moral norms*, this study does not take into account some moral issues which, though studied in literature, were not included in the pilot study concerning organic products, for example political motivations associated to the development of local economies, or to animal welfare, such as avoiding the employment of cruel breeding methods, drugs or hormones, which were not mentioned in the pilot study. Considering these limitations, it could be interesting to carry out the same analysis using different questions and methods of data collection that could provide a better understanding of purchase behavior and the consumption experience of organic products, in order to study the link between intention to buy and purchase behavior as well. The majority of studies on organic products, the present one included, are questionnaire-based surveys or personal interview surveys (for reviews, Hughner *et al.* 2007; Thompson 1998), that enable researchers to collect a larger number of data investing a small amount of resources. Other methods, such as a *focus group* (for example, Hill and Lynchehaun 2002; Padel and Foster 2005) and analyses of scanned data (for example, ACNielsen Homescan) have rarely been employed.

Notwithstanding these limitations, this study contributes to shed light on determinants of *intention* to buy organic food products. It stresses that ethical consumers feel morally obliged to protect their own and their relatives' health, when buying natural products. At the same time, they can perceive themselves as better individuals, people who care of their wellness and of that of their loved ones, and who safeguard their lives from useless and exaggerated risks.

Gianluigi Guido, M. Irene Prete, R. Christian Maloumby-Baka
and Carolina Buffa

3. Effects of product personality, moral norms and moral disengagement on the intention to buy organic food products[1]

3.1. Introduction

Over the past several years, research has identified a category of consumers defined as "ethical consumers", those who are interested in a large number of principled topics, from the safeguard of the environment to the protection of health, human rights and the well-being of animals. Nowadays, the market of ethical consumption has enormous potential and has for some time shown high rates of growth (Leatherhead Food International 2007). Ethical consumption includes, among the others, the organic food market, which, though being a niche within the food sector in many national markets, has nonetheless seen strong growth in the last few years – e.g., the global organic market rose by 13.6% in 2006 (Datamonitor 2007). Yet, despite the great potential of this market, there is still no deep understanding of the mechanisms at the basis of ethical consumers' decision making. As regards organic agriculture, in particular, some researcher has studied the connection between moral considerations and the purchase of organic food products by applying prevalently Schwartz's (1992) *Value theory*, or Ajzen's (1991) *Theory of planned behavior*.

---

1 The Authors would like to thank Dr Alessandro M. Peluso for his comments on a previous version of this manuscript and Giorgia A. Pollara for her support in data collection and codification.

The objective of the present study is to broaden the analysis of how significant this ethical dimension can be in the purchasing intention of organic food products, as well as to propose an intercultural comparison between Italy and France as regards components of the organic food product image – identified through the *Five-factor model* of personality (cf. Digman 1990) – which is supposed to have an impact on these ethical determinants. In order to evaluate how people's ethics influences purchase intention, the *Theory of planned behavior* has been extended here to two factors which identify the ethical component: *Moral norms*, which represent personal beliefs regarding what is right or wrong (Parker, Manstead and Stradling 1995), and *Moral disengagement* (Bandura 1986, 1999), which refers to those cognitive mechanisms which allow individuals' peace of mind when their behavior infringes their own moral codes. Until now, the construct of *moral disengagement* has been primarily employed in situations of violence and aggressiveness. In the present study, it is applied to normal behavior – in accordance with Caprara *et al.* (2006). The study is structured as follows: Section 2, after presenting organic agriculture and other cultivation methods, reviews the literature regarding the demographic and psychographic profile of the organic food product consumers and presents the *Prospect method* (Caprara and Barbaranelli 2000; Caprara, Barbaranelli and Guido 2000) – which integrates the *Theory of planned behavior* and the *Five-factor model* within a single model – by tracing, specifically, the moral antecedents in the purchase intention of organic food products. Section 3 introduces the research objectives, methodology and results; whereas, Section 4 synthesizes findings and discusses implications, limitations, and future lines for research.

3.2. Determinants of intention to buy organic food products

Organic agriculture is a type of production which is natural, not genetically manipulated, and which "maintains and replenishes soil fertility without the use of toxic and persistent pesticides and fertilizers" (Organic Trade Association 2008: 1). It is an agricultural system which produces healthy food free of chemical substances, in full respect of the environment, and which differentiates itself from other methods of cultivation such as: (*i*) *conventional agriculture*, which employs crop intensive technologies aimed at maximizing the volumes of production and profit and which makes an unrestricted use of chemical synthetic substances for fertilizing and insecticide purposes; and (*ii*) *biotechnological agriculture*, which applies recent scientific achievements, including those in the field of genetic engineering, in order to modify plants, animals and micro-organisms, and to promote improvements in production and agronomy (United States Department of Agriculture 2008).

3.2.1. The consumer of organic food products

Several studies have researched into the demand for organic food products in order to identify the demographic and psychographic profile of the standard consumer (see Hughner *et al.* 2007, for a review). As regards *demographic characteristics*, most of the studies have considered gender, marital status, family size, income and level of education. Generally, consumers of organic food products are women (Buzby and Skees 1994; Byrne, Bacon and Toensmeyer 1994; Davies, Titterington and Cochrane 1995; Govindasamy and Italia 1997; Groff, Kreidor and Toensmeyer 1993; O'Donovan and McCarthy 2002) and, in particular, middle-aged women (Aguirre 2007), even though some studies (Buzby and Skees 1994; Magnusson *et al.* 2001) have revealed the young segment to possess a higher propensity towards such products. Nonetheless middle-aged consumers are the most frequent purchasers, this being explained, according to Hughner *et al.* (2007), by the higher prices of organic food compared to those of conventional products which means that only certain consumers can afford the former: in fact, the greater financial resources of middle-aged consumers allow them to pay a higher premium price for organic food products. Research findings have also indicated that consumers of organic food products live mainly in the family home (Aguirre 2007) together with their children (Wier and Smed 2000). Although there is no general consensus, many authors agree that both an upper-middle income and a relatively high level of education are important characteristics of such standard consumers (Aguirre 2007; Chinnici, D'Amico and Pecorino 2002; Davies, Titterington and Cochrane 1995; Fotopoulos and Krystallis 2002; O'Donovan and McCarthy 2002).

Some studies have also focused on understanding the organic food product consumer's *psychographic profile*, in the sense of personality and lifestyle. The purchase of organic food products is part of a wider lifestyle choice, deriving from a particular value system, which influences personality, attitudes and consumer behavior (Schifferstein and Oude Ophuis 1998). Such values have been studied by various authors (e.g., Dreezens *et al.* 2005; Makatouni 2002; Zanoli and Naspetti 2002) who have emphasized how regular organic food product consumers draw inspiration from a set of values, such as: *Universalism*, i.e. the understanding, tolerance and protection of all people and of nature; *Altruism*, i.e. the generosity in relationships with others; *Ecology*, i.e. the harmony with the universe and with a sustainable future; *Benevolence*, i.e. the improvement of the well-being of the people with whom one is in contact; and *Spirituality*, i.e. the unity with nature. Furthermore, the organic food product consumer leads an "alternative" lifestyle, being often a vegetarian, an active environmentalist and a user of alternative medicine (Cicia, Del Giudice and Scarpa 2002).

3.2.2. The Prospect method

Products, in general, and organic food products, in particular, possess not merely a functional value but also a symbolic meaning, which can be personified through their product image. The *Prospect method* (Caprara and Barbaranelli 2000; Caprara, Barbaranelli and Guido 2000) is a framework employed in the present study in order to evaluate the influence that the traits of *product image* – in this case, of the image of organic food products, operazionalized through the construct of *product personality* – exerts on the determinants of purchase *intention*. This method combines two models that are well-established in literature: the *Five factors model* of human personality (a.k.a. the *Big*

Five factors model, see Digman 1990, for a review), which allows researchers to find the fundamental features of brands and products when applied in a marketing setting; and the *Theory of planned behavior* by Ajzen (1991), which identifies the main determinants of consumers' purchase intention, assumed as the most reliable predictor of purchasing behavior. The *Prospect method* integrates, in a single research tool, both models with the goal of verifying whether and to what extent latent dimensions of brand and product image (as assessed by the *Five-factor model*) influence purchase intention and its cognitive determinants as measured by the *Theory of planned behavior* (see Figure 3.1). In this way, the *Prospect method* allows the identification of: (*i*) the personality elements of brands and products which can influence the main determinants of purchase intention; and (*ii*) the most appropriate adjectives to be used in marketing communications, among those that are the best in eliciting in consumers' minds the traits of product image which can have an impact on the determinants of purchase intention.

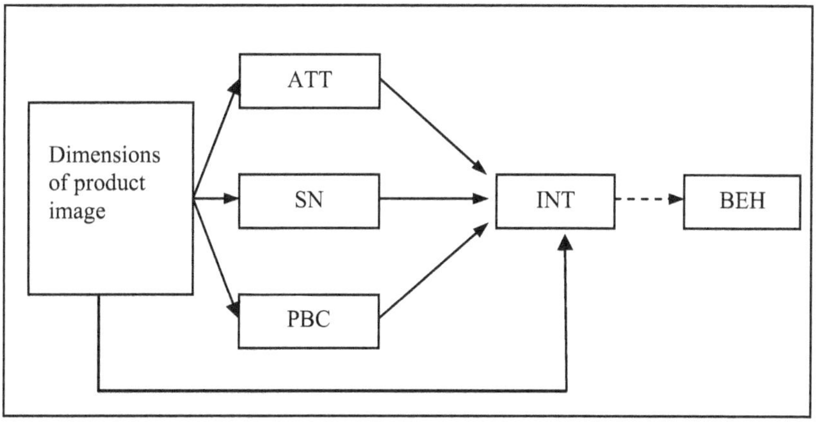

Note: Dimensions of product image = Agreeableness, Conscientiousness, Emotional stability/Neuroticism, Extroversion/Introversion, Openness to experience; ATT = Attitude; SN = Subjective norm; PBC = Perceived behavioral control; INT = Intention to perform a specific behavior; BEH = Behavior (e.g., the intention to buy organic food products).

Figure 3.1. The influence model according to the *Prospect method.*

3.2.2.1. The Five-factor model of personality

Through the application of factor analysis on a potentially infinite set of adjectives describing persons, the *Five-factor model* (hereafter FFM) of personality (cf. Digman 1990) identifies as many principle dimensions (the so-called *Big Five factors*) that describe the basic latent traits of human personality. According to some authors (e.g., Aaker 1997; Plummer 2000), since branded products can also have a personality, it is possible to employ the FFM to identify also the latent dimensions of products and brand images, even though some allowances need to be made (Caprara and Barbaranelli 1996; Caprara, Barbaranelli and Guido 1998, 2000, 2001). Although there is a lack of agreement in literature on the labels to be attributed to the five factors, the most common are: *Agreeableness*, referring to the importance given to information and the opinions of others regarding the correctness of one's behavior: *Conscientiousness*, referring to the degree of meticulousness and preciseness in which an individual aims at specific objectives; *Emotional stability/Neuroticism*, referring to the degree of patience with which an individual reacts to external stimuli; *Extroversion/Introversion* regarding the number of interpersonal relations an individual feels at ease with; and *Openness to experience*, referring to the level of open-mindedness an individual shows towards new ideas and experiences.

A search through the available sources reveals no study applying the FFM to organic food products. Given the potential of this market and the importance of image in the purchasing of food, the present study aims therefore to analyze the possible latent dimensions of the image of organic food products in two European countries (i.e., Italy and France) in order to evaluate how far the differences between the two cultures can explain the different perceptions consumers may have of the products under examination.

3.2.2.2. The Theory of planned behavior

The *Theory of planned behavior* (hereafter TPB) is a predictive model based on the fundamental assumption that the *Intention* to perform (or not) a specific *Behavior* (e.g., buying) is the best predictor of that *behavior*. In its turn, *intention* is a function of a series of determinants whose importance varies according to the target behavior. They are (Ajzen 1991; Ajzen and Fishbein 1980): *Attitude*, defined as the level of favorable or unfavorable propensity of an individual towards the behavior considered; *Subjective norm*, defined as the level of social pressure which people who are taken as important exert on an individual to make him/her adopt a specific form of behavior; and *Perceived behavioral control*, defined as the perception by an individual of the degree of ease or difficulty required to adopt a specific form of behavior.

The TPB has been also extended to include "additional predictors if it can be shown that they capture a significant proportion of the variance in intention or behavior" (Ajzen 1991: 199). Research has generally taken into account (cf. Eagly and Chaiken 1993): *Past behavior*, considered as a good predictor of future behavior, provided that the internal and external factors that have determined it remain stable over time (Jaccard and Blanton 2005); *Self-identity*, defined as the set of durable characteristics which people assign to themselves (see, for example, Biddle, Bank and Slavings 1987); and *Moral norms*, defined as personal beliefs regarding what is right or wrong (Parker, Manstead and Stradling 1995).

The purchase intention of organic food products has been analyzed in various studies that have tried to determine which variable is more important in influencing *intention*. Generally, a positive *attitude* displayed by consumers towards organic food products has been shown to be very influential in most studies (Michaelidou and Hassan 2008; Sparks and Shepherd 1992; Thøgersen 2002). On the contrary, different authors have drawn different conclusions as regards the influence of *subjective norm*: according to some of them, *subjective norm* has a great influence, both direct (Sparks and Shepherd 1992) and indirect (Tarkiainen and Sundqvist 2005), via *attitude*; whereas Thøgersen (2002) has shown that *subjective norm* is not a significant predictor of *intention*. Also *perceived behavioral control* – which in theory seems to negatively influence *intention* due to major obstacles perceived during purchase, such as price and product availability (Sparks and Shepherd 1992) – does not emerge as a significant variable in Tarkiainen and Sundqvist's study (2005). The extension of the TPB to *past behavior*, though used only in Thøgersen's (2002) study, has significantly improved the predictive capability of the model applied to the case of organic food products. *Self-identity* has instead been included in several studies (e.g., Michaelidou and Hassan 2008; Sparks and Shepherd 1992), which claim that its influence on *intention* of purchasing organic food products is direct and significant; Michaelidou and Hassan (2008) believe this influence to be both direct and mediated by *attitude*. Also as regards the inclusion of *moral norms* – considered necessary not only for behaviors with a high ethical content (Conner and Armitage 1998; Manstead 1999), but also in low involvement situations (such as the buying of food) – there are two branches of research concerning their influence on the purchase *intention* of organic food products: some researchers consider their impact as direct and significant (Thøgersen 2002); others consider it as indirect, taking *moral norms* as an antecedent of *attitude* (Dreezens *et al.* 2005; Honkanen, Verplanken and Olsen 2006).

3.2.3. Moral antecedents in the purchasing of organic food products

Moral considerations are often considered important by consumers of organic food products. Such considerations are one of the topics of this study. The purchase of organic food products, in fact, implies ethical evaluations associated with consumers' health protection (Chinnici, D'Amico and Pecorino 2002; Lockie *et al.* 2002; Makatouni 2002; Padel and Foster 2005; Zanoli and Naspetti 2002), moral responsibility towards the users of these products (Finch 2005; Makatouni 2002; Raats, Shepherd and Sparks 1995), the well-being of animals (Honkanen, Verplanken and Olsen 2006; Makatouni 2002; McEachern and McClean 2002; Padel and Foster 2005); and political aspects such as the promotion of the local economy and the support of commerce, as well as the preference for goods from countries with democratic regimes (Honkanen, Verplanken and Olsen 2006). In order to understand the role played by moral dimensions in the purchase of organic food products, the present study extends the TPB by including *moral disengagement* (Bandura 1986, 1990, 1999) as a further determinant of the purchase *intention*, in addition to the classic *moral norms*. The extension of the TPB to these two elements is justified by the fact that *moral disengagement* and *moral norms* are not necessarily opposite concepts, hence they can both influence *intention*: i.e., *moral norms* positively, whereas *moral disengagement* negatively. The present study hypothesizes that consumers of organic food products are influenced by moral obligations and that, at the same time, they are involved in tasks of an ethical nature.

3.2.3.1. Moral disengagement

The construct of *moral disengagement*, developed by Bandura (1999) within the context of social-cognitive theory, represents "the process by which an individual proactively suspends his/her moral standards in an effort to reduce the self-censure that ordinarily accompanies conduct that violates one's moral code" (Cava 2000: 492). The conduct of an individual is, in fact, bound by a moral code developed over time and "in this process of auto-regulation the individual abstains from behaving in ways which would break his/her moral code, since such behaviors would result in self-punishment" (Bandura 1999: 193–194). The reduction of self-punishment following the process of *moral disengagement* provides an individual with the freedom to break his/her moral code.

Various mechanisms allow the individual to selectively suspend his/her moral principles in order to perform acts that would otherwise trigger self-punishment. Bandura (1986) has identified eight of them: *Moral justification*, when a reproachful conduct becomes personally and socially acceptable by appealing to moral goals; *Euphemistic labeling*, when metaphorical language is used to make offences and atrocities genteel and tolerable; *Advantageous comparison*, when a reprehensible action is made less blameworthy in relationship with even more reprehensible actions; *Displacement of responsibility* to other individuals, when people blame superiors, usually those in authority, for their harmful actions; *Diffusion of responsibility*, when individual liability for reprehensible actions is attenuated by assigning it to a group of people; *Negligence or distortion of consequences*, when the effects of a harmful action are minimized or ignored; *Dehumanization*, when a victim is deprived of his/her individual dignity in order to avoid any empathic reaction; and *Attribution of blame*, when the guilt of adversaries or circumstances allows denial of one's responsibility for an offence.

Although Bandura (1986) has studied *moral disengagement* with special reference to aggressive and violent forms of behavior, the use of this concept in the present study is supported by Caprara *et al.* (2006), who demonstrate how this construct can be applied also to normal forms of behavior, such as "minor" transgressions, as is the case of not choosing organic food products, thus exposing, for example, to risk one's own health and that of one's loved ones.

3.3. Methodology

3.3.1. Research objectives

The present study has two objectives: firstly, to analyze the purchase behavior of organic food consumers by means of Ajzen's (1991) TPB, which is extended in order to consider the importance of moral dimensions; secondly, to evaluate – through the *Prospect method* (Caprara, Barbaranelli and Guido 2000) – the influence that the organic products' personality traits exert on determinants of consumers' purchase *intention* (moral antecedents included).

As regards the first objective, *moral norms* and *moral disengagement* have been chosen as possible additional predictors of purchase *intention* – together with *attitude*, *subjective norm* and *perceived behavioral control* – in order to evaluate the influence of the moral dimensions in the TPB, when applied to organic food products. While the inclusion of *moral norms* to explain the purchase behavior of organic products has been the topic of previous studies (Honkanen, Verplanken and Olsen 2006; Thøgersen 2002), no research has so far extended the TPB to *moral disengagement*, and its use in the present study is justified by extension from situations that imply major transgressions, such as violent behaviors, to more ordinary situations, such as not buying products which can have positive effects both socially and personally (Caprara *et al.* 1996, 2006).

In particular, *moral norms* have been measured considering salient beliefs, i.e. those beliefs easily accessible to memory (Fishbein and Ajzen 1975), such as the protection of one's health and that of one's loved ones, whereas *moral disengagement* (Bandura 1986) has been measured through the *Moral disengagement scale* (Caprara *et al.* 1996), consisting of 76 items, grouped into eight categories referred to the same number of *moral disengagement* mechanisms. The reason for the inclusion of both constructs into the TPB is – as mentioned above – that *moral norms* and *moral disengagement* are not necessarily opposed, since they consist of different elements which can be activated by contextual factors: social factors, such as the presence of other people; and situational factors (e.g., environmental ones). In other words, even a person with a moral sense might behave in a morally disengaged way in particular situations and contexts; social factors do not exclude the situational ones, and they can be both significant predictors of *intention*.

The basic assumption of the present study is that organic food product consumers are influenced by moral considerations and that they show a lesser degree of disengagement with regard to ethical topics. From an operational point of view, the aim of this study is to demonstrate that *moral norms* directly influence the purchase *intention* of organic food products, while *moral disengagement* can have a statistically significant, negative effect on *intention*.

As far as the second research objective is considered, the aim here is to verify which aspects of organic food product image – operazionalized through product personality – mostly affect the significant determinants of the purchase *intention*. Moreover, this research has been extended to two European countries – namely, Italy and France – in order to evaluate whether cultural differences play a role in this respect. The choice of these two countries is justified by the fact that the goal is to compare the Italian results with those relative to a more advanced country (France) as far as ethical consumerism goes (Datamonitor 2005). In fact, since values and moral standards are generally shared within the same culture, it can be useful to explain potential differences in the behavior of citizens of different nationalities by considering the similarities and the differences between the relevant cultures.

3.3.2. Procedure

The research has been divided into two steps: a *pilot study*, designed to identify the salient beliefs at the basis of *attitude*, *subjective norm*, *perceived behavioral control*; and a *main study*, designed to evaluate the weight of these variables, *moral norms* and *moral disengagement* – measured through the *Moral disengagement scale* (Caprara *et al.* 1996) – on the *intention* to buy. Finally, the *Prospect method* (Caprara, Barbaranelli and Guido 2000) was employed to evaluate the influence of organic food product personality traits on determinants of the purchase *intention*.

3.3.2.1. The pilot study

The pilot study, designed to identify the salient beliefs needed for the main survey, was conducted on a sample of 40 subjects – 20 males and 20 females, grouped into five age-groups (15-29; 30-41; 42-53; 54-65; 66+ years old) – selected through the direct interception technique (i.e., the random interview of consumers at the end of their purchase act) in specialized shops.

Following a common practice in testing the TPB (Ajzen and Fishbein 1980; Fishbein and Ajzen 1975), an open-ended questionnaire was administered to this sample in order to identify the salient beliefs at the basis of *intention* determinants, namely: *attitude*, by asking them to enumerate the advantages and disadvantages of their purchase of organic and non-organic food products; *subjective norm*, by asking them to specify possible people or groups of people that might approve or disapprove of their purchase of such products; *perceived behavioral control*, by asking them to enumerate facts or circumstances that could facilitate or impede the purchase *intention*; and *moral norms*, by asking two questions aimed to specify possible ethical reasons behind their purchase *intention*. Further questions were introduced to investigate: the frequency and the likelihood of purchase *intention* during the two weeks following the test; and the adjectives that best describe the organic food product image, to be used to find any further latent factor, not included among the *Big Five* ones (so-called *non-Big Five* factor). The results of the pilot study, obtained by analyzing the protocols of answers and extracting the most frequent ones, allowed the construction of the main close-ended questionnaire.

3.3.2.2. The main study

The main study was carried out both in Italy and in France on a randomly selected sample of 212 subjects, 160 Italians and 52 French (see Table 3.1, for a description of the sample). A closed-ended questionnaire was administered to the sample; it was made up of four sections: a first section, containing a series of items, defined on a 7-point Likert scale designed to measure the probability and the importance of the elements of the TPB (i.e., *attitude*, *subjective norm*, *perceived behavioral control*, and purchase *intention*), as well as the probability of two items regarding *moral norms*; a second section, structured in such a way as to identify the image of organic food products through the use of 20 adjectives – 15 Big Five attributes, chosen among those marker attributes, as indicated by Caprara and Barbaranelli (2000), that are best able to elicit the five main dimensions of personality (three adjectives for each Big Five factor), and 5 non-Big Five attributes, namely *healthy, tasty, genuine, natural* and *good* – evaluated through a 7-point semantic differential scale, selected from those most frequently named in the pilot stage; a third section, regarding social-demographic

Countries	*N*	*Gender*	*Age*
Italy	160	M = 80 ss. F = 80 ss.	40 ss. (16-29 y.o.) 40 ss. (30-39 y.o.) 40 ss. (40-49 y.o.) 40 ss. (50-59 y.o.)
France	52	M = 21 ss. F = 31 ss.	30 ss. (16-29 y.o.) 10 ss. (30-39 y.o.) 7 ss. (40-49 y.o.) 5 ss. (50-59 y.o.)

Table 3.1. Sampling characteristics in the main study.

data (i.e., gender, age, frequency and point-of-purchase of food products); and a fourth section, regarding the *Moral disengagement scale* (Caprara *et al.* 1996) consisting of 76 items, defined on a 5-point Likert scale. In order to consider *moral disengagement* as an additional determinant of the extended TPB, the mean value of the 76 items was considered in data analysis.

3.4. Results

3.4.1. Determinants of the purchase intention of organic food products

In order to evaluate whether *attitude*, *subjective norm*, *perceived behavioral control*, *moral disengagement* and *moral norms* can be considered valid predictors of organic food product purchase *intention*, a multiple linear regression analysis was employed, considering *intention* as the dependent variable and all the determinants of the extended TPB as independent variables.

The results of the first part of the study, conducted in Italy (see Table 3.2, *below*), show that the model adapts well to real data (R^2 = .562), since the total variance of *intention* explained is over 50%, in line with results for this kind of study (Sutton 1998). On the basis of beta values, it can be seen that the determinants which influence significantly the purchase *intention* of organic food products are: *perceived behavioral control*, which shows the highest beta value (β = .376, $p < .001$); *moral norms* (β = .309, $p < .001$); *attitude* (β = .261, $p < .001$); *subjective norm* (β = .233, $p < .001$); whereas *moral disengagement* does not significantly influence *intention*, since the probability was not low enough ($p > .05$), even though – as hypothesized – a negative beta value does occur. The non-significant influence of *moral disengagement* can be probably explained on the basis of the procedure used to measure this construct in the first place, which makes the respondents sensitive to the "social desirability bias" – the

Variables	*B*	*Beta*	*t*	*p*
Constant	-26.875		-5.064	.000
ATT	*.109*	*.261*	*4.472*	*.000*
SN	*.089*	*.233*	*3.559*	*.000*
PBC	*.160*	*.376*	*6.641*	*.000*
MD	-1.426	-.050	-.909	.365
MN	*1.083*	*.309*	*4.553*	*.000*

Note: Dependent variable: Intention; $N = 160$; $R = .749$; $R^2 = .562$; Adjusted $R^2 = .548$. ATT = Attitude; SN = Subjective norm; PBC = Perceived behavioral control; MD = Moral disengagement; MN = Moral norms.

Table 3.2. Regression results for the intention to buy organic food products in Italy.

tendency of interviewees to give replies that will be evaluated positively by others (Fisher 1993).

In order to understand which salient beliefs – at the basis of the variables that significantly influence *intention* – have the most weight, the mean (M) and standard deviation (SD) values were calculated. From the analysis of the results shown in Table 3.3 and Tables 3.4–3.6 (*below*), it is possible to infer that, as regards the organic food products in Italy, the most salient beliefs are: for *perceived behavioral control*, that it improves health (M = 30.013, SD = 11.917) and that stores can offer a great selection (M = 23.056, SD = 10.612); for *moral norms*, health protection (M = 4.563, SD = 1.899); for *attitude*, following a healthy and balanced diet (M = 25.475, SD = 10.199) and eating completely natural products (M = 24.663, SD = 9.071); and for

Control beliefs	*M*	*SD*
Benefits to health	30.013	11.917
High variety/assortment	23.056	10.612
High availability in stores	22.163	10.229
Distance and difficulty of reaching stores	19.769	9.965
Low prices	18.331	10.468

Table 3.3. Control beliefs affecting *perceived behavioral control* for organic food products in Italy.

Relevant beliefs behind moral norms	*M*	*SD*
I feel obliged to purchase organic food to safeguard my health	4.563	1.899
I feel obliged to purchase organic food to safeguard the health of my family	4.506	2.009

Table 3.4. Salient beliefs affecting *moral norms* for organic food products in Italy.

Behavioral beliefs	*M*	*SD*
Following a healthy and balanced diet	25.475	10.199
Eating completely natural products	24.663	9.071
Avoiding possible food poisoning and/or allergies	23.813	9.677
High prices	23.238	10.154
Difficulties in finding specialized shops	21.413	9.431

Table 3.5. Behavioral beliefs affecting *attitude* for organic food products in Italy.

Normative beliefs	*M*	*SD*
Partner's (wife/husband, fiancée/fiancé) approval	26.413	15.652
Family's approval	24.438	14.908
Friends' approval	14.413	11.051

Table 3.6. Normative beliefs affecting *subjective norm* for organic food products in Italy.

subjective norm, the social pressure exercised by one's partner (M = 26.413, SD = 15.652) and by the family (M = 24.438, SD = 14.908).

The same analysis was carried out, in the second part of the study, in reference to the purchase *intention* of organic food products in France (see Table 3.7, *below*). Although it shows a low level of adaptation to real data (R^2 = .302), it is possible to infer that *moral norms* (β = .414, p < .01) turned out to be, for French consumers, the variable that influenced *intention* most – followed by *perceived behavioral control* (β = .282, p < .05).

Variables	*B*	*Beta*	*t*	*p*
Constant	-15.952		-1.064	.293
ATT	.083	.171	1.264	.213
SN	-.003	-.007	-.050	.960
PBC	*.095*	*.282*	*2.250*	*.029*
MD	-.538	-.015	-.105	.917
MN	*1.539*	*.414*	*2.901*	*.006*

Note: Dependent variable: Intention; $N = 52$; $R = .550$; $R^2 = .302$; Adjusted $R^2 = .226$. ATT = Attitude; SN = Subjective norm; PBC = Perceived behavioral control; MD = Moral disengagement; MN = Moral norms.

Table 3.7. Regression results for the intention to buy organic food products in France.

From the analysis of the mostly salient beliefs at the basis of the variables which significantly influence *intention* (see Tables 3.8–3.9), it is possible to affirm that those who buy organic food products in France are influenced mainly by consumers feeling compelled to purchase organic food to safeguard their health (M = 4.000, SD = 1.990), as well to obtain benefits from it (M = 37.404, SD = 10.008).

Relevant beliefs behind moral norms	*M*	*SD*
I feel obliged to purchase organic food to safeguard my health	4.000	1.990
I feel obliged to purchase organic food to safeguard the health of my family	3.981	2.063

Table 3.8. Salient beliefs affecting *moral norms* for organic food products in France.

Control beliefs	*M*	*SD*
Benefits to health	37.404	10.008
High variety/assortment	24.635	12.637
High availability in stores	22.942	11.863
Distance and difficulty of reaching stores	21.750	13.171
Low prices	20.058	14.056

Table 3.9. Control beliefs affecting *perceived behavioral control* for organic food products in France.

3.4.2. Results of the moral disengagement scale

From the analysis of the *moral disengagement* items' average values it can be inferred that, in Italy, consumers of organic food products disagree with statements like: "It doesn't make sense for an individual to care about environmental damage, since harmful effects are produced at a collective level" (M = 1.42, SD = .82), or "Companies that produce things of public interest shouldn't be prosecuted if they pollute the environment" (M = 1.48, SD = .98). In France, too, consumers declared that they disagree with the same statements – "It doesn't make sense for an individual to care about environmental damage, since harmful effects are produced at a collective level" (M = 1.38, SD = .72), and "Companies that produce things of public interest shouldn't be prosecuted if they pollute the environment" (M = 1.54, SD = .83) – but also with: "It doesn't make sense to care about those who smoke since the damage they do is almost irrelevant compared to that related to environmental pollution" (M = 1.63, SD = .86).

Yet, it has to be emphasized how results in general display low-medium values (always below 4): the average value of *moral disengagement* for Italian consumers of organic food products equals, in fact, 2.06; while, in France, it equals 2.02. Respondents hence declared themselves to be "Not morally disengaged" (that is, in a sense, ethically involved). This result deserves some reflection as it raises issues about the reliability of answers. In fact, while the literature recognizes the effectiveness of indirect measurement of elements related to the intimate and personal sphere, the scale used measures *moral disengagement* through the eliciting of an explicit opinion about the 76 situations considered. Such direct *moral disengagement* measurement can explain one tendency shown in results: it can, in fact, act as a barrier to the freedom of expression of people who will hardly admit that they would behave or would agree with behaviors conflicting with common moral codes when asked in public.

3.4.3. The Prospect method

The *Prospect method* (Caprara and Barbaranelli 2000; Caprara, Barbaranelli and Guido 2000) was used to evaluate the influence of organic food product personality traits on determinants of the purchase *intention*. By integrating the TPB and the FFM, this statistical method allows one to identify the latent dimensions of organic food product image, and after that to regress, one by one, the variables of the extended TPB – *attitude, subjective norm, perceived behavioral control, moral disengagement* and *moral norms* – on the six traits which were found (i.e., the Big Five factors, i.e., *Agreeableness, Conscientiousness, Emotional stability/Neuroticism, Extroversion/Introversion*, and *Openness to experience*, plus a non-Big Five factor, which was named *Naturalness* after the non-Big Five attributes saturated on it).

3.4.3.1. The latent dimensions of organic food product image

In order to identify the latent dimensions of organic food product image in Italy and France, factorial analyses were employed by applying the extraction method of the principal components, with an orthogonal Varimax rotation, and by imposing a solution with six factors, five of which were found to be the Big Five ones, and one representing the non-Big Five factor. To identify each factor, the saturation levels of each "marker" attribute – that is of every adjective able to elicit, better than others, the underlying factors – was observed, considering a level of acceptability equal to .40. Table 3.10 (*below*) shows the results related to the latent dimensions of organic food product images in the two countries.

In Italy, the main dimension of organic food product image is *Extroversion*, saturated by adjectives like *tasty, good, cheerful* and *energetic*, followed by *Naturalness*, saturated by adjectives like *genuine, natural*, and *healthy*. The six factors explain 71.44% of the total variance, which demonstrates the high level of representativeness of such dimensions. In France, results show that *Naturalness*, saturated by adjectives like *reliable, genuine, safe, healthy*, and *natural*, is the principal dimension of organic food product image, followed by

*ITALY**		*FRANCE***	
Factors	*Saturations*	*Factors*	*Saturations*
I. Extroversion		*I. Naturalness*	
Tasty	.764	Reliable	.755
Good	.719	Genuine	.737
Cheerful	.640	Safe	.729
Energetic	.581	Healthy	.632
		Natural	.629
II. Naturalness		*II. Conscientiousness*	
Genuine	.897	Innovative	.813
Natural	.854	Funny	.792
Healthy	.834	Diligent	.627
		Careful	.627
		Friendly	.586
III. Conscientiousness		*III. Agreeableness*	
Reliable	.764	Cheerful	.772
Careful	.698	Warm	.700
Safe	.600	Affectionate	.688
Competitive	.584	Energetic	.607
IV. Agreeableness		*IV. Emotional Stability*	
Affectionate	.790	Calm	.814
Warm	.743	Tolerant	.657
Friendly	.717	Stable	.613
V. Emotional Stability		*V. Openness to experience*	
Tolerant	.804		
Stable	.668	Tasty	.809
Calm	.616	Good	.636
Diligent	.512		
VI. Openness to experience		*VI. Extroversion*	
		Competitive	.771
Innovative	.903		
Funny	.449		

Note: Processing method: analysis of the principal components; Rotation method: Varimax rotation with Kaiser normalization; * = Convergence with 19 iterations; ** = Convergence with 12 iterations.

Table 3.10. Dimensions of organic food product images in Italy and in France.

Conscientiousness, saturated by adjectives like *innovative*, *funny*, *diligent*, *careful*, and *friendly*. The total variance explained was slightly better than in Italy and equal to 72.40%, with a loss of less than one third of the information.

3.4.3.2. *The influence of the image dimensions on the determinants of intention*

A standard regression analysis was carried out to examine the influence of each of the six latent dimensions of organic food product images, both in Italy and in France, exerted on the determinants of purchase *intention* that were revealed to be significant – as well as, directly, on purchase intentions.

Results related to organic food product image in Italy (see Table 3.11 and Tables 3.12–3.15, *below*) show, in order of relevance, that the most significant image traits connected to *perceived behavioral control* are *Naturalness* ($\beta = .249, p < .001$), and *Conscientiousness* ($\beta = .192, p < .05$); whereas the only significant trait to influence *moral norms* is *Extroversion* ($\beta = .471, p < .001$). *Agreeableness* ($\beta = .246, p < .001$) and *Emotional Stability* ($\beta = .169, p < .05$) are the traits influencing *attitude*. *Extroversion*, again, is the latent image trait that both significantly influences *subjective norm* ($\beta = .312, p < .001$), and directly *intention* ($\beta = .408, p < .001$) – in this latter case, together with *Conscientiousness* ($\beta = .168, p < .05$). As far as France is concerned (see Tables 3.16–3.17, *below*), the factors that mostly influence *moral norms*

Variables	*B*	*Beta*	*t*	*p*
Constant	113.331		48.382	.000
Extroversion	3.865	.123	1.645	.102
Naturalness	*7.791*	*.249*	*3.316*	*.001*
Conscientiousness	*6.007*	*.192*	*2.556*	*.012*
Agreeableness	3.551	.113	1.511	.133
Emotional stability	-.091	-.003	-.039	.969
Openness to experience	3.616	.115	1.539	.126

Note: Dependent variable: Perceived behavioral control; $N = 160$; $R = .374$; $R^2 = .140$; Adjusted $R^2 = .106$.

Table 3.11. Regression of *perceived behavioral control* on the image factors of organic food products in Italy.

Variables	*B*	*Beta*	*t*	*p*
Constant	9.069		34.309	.000
Extroversion	*1.790*	*.471*	*6.749*	*.000*
Naturalness	.452	.119	1.703	.091
Conscientiousness	.471	.124	1.778	.077
Agreeableness	.102	.027	.385	.701
Emotional stability	-.161	-.042	-.609	.543
Openness to experience	.179	.047	.676	.500

Note: Dependent variable: Moral norms; $N = 160$; $R = .506$; $R^2 = .256$; Adjusted $R^2 = .227$.

Table 3.12. Regression of *moral norms* on the image factors of organic food products in Italy.

Variables	*B*	*Beta*	*t*	*p*
Constant	118.600		48.518	.000
Extroversion	.933	.029	.380	.704
Naturalness	.533	.017	.218	.828
Conscientiousness	.384	.012	.156	.876
Agreeableness	*7.888*	*.246*	*3.217*	*.002*
Emotional stability	*5.410*	*.169*	*2.206*	*.029*
Openness to experience	3.451	.108	1.407	.161

Note: Dependent variable: Attitude; $N = 160$; $R = .320$; $R^2 = .102$; Adjusted $R^2 = .067$.

Table 3.13. Regression of *attitude* on the image factors of organic food products in Italy.

Variables	*B*	*Beta*	*t*	*p*
Constant	65.263		24.973	.000
Extroversion	*10.912*	*.312*	*4.162*	*.000*
Naturalness	2.677	.077	1.021	.309
Conscientiousness	4.502	.129	1.717	.088
Agreeableness	.010	.000	.004	.997
Emotional stability	-4.327	-.124	-1.650	.101
Openness to experience	-1.718	-.049	-.655	.513

Note: Dependent variable: Subjective norm; $N = 160$; $R = .371$; $R^2 = .138$; Adjusted $R^2 = .104$.

Table 3.14. Regression of *subjective norm* on the image factors of organic food products in Italy.

Variables	*B*	*Beta*	*t*	*p*
Constant	16.863		17.907	.000
Extroversion	*5.435*	*.408*	*5.754*	*.000*
Naturalness	1.294	.097	1.370	.173
Conscientiousness	*2.245*	*.168*	*2.377*	*.019*
Agreeableness	1.853	.139	1.961	.052
Emotional stability	.133	.010	.140	.889
Openness to experience	1.190	.089	1.259	.210

Note: Dependent variable: Intention; $N = 160$; $R = .481$; $R^2 = .231$; Adjusted $R^2 = .201$.

Table 3.15. Regression of purchase *intention* on the image factors of organic food products in Italy.

Variables	*B*	*Beta*	*t*	*p*
Constant	7.981		16.533	.000
Extroversion	-.234	-.061	-.479	.634
Naturalness	*1.384*	*.362*	*2.838*	*.007*
Conscientiousness	-.200	-.052	-.411	.683
Agreeableness	*1.320*	*.345*	*2.709*	*.010*
Emotional stability	-.319	-.084	-.655	.516
Openness to experience	.249	.065	.510	.612

Note: Dependent variable: Moral norms; ; $N = 52$; $R = .518$; $R^2 = .268$; Adjusted $R^2 = .170$.

Table 3.16. Regression of *moral norms* on the image factors of organic food products in France.

Variables	*B*	*Beta*	*t*	*p*
Constant	20.106		13.137	.000
Extroversion	2.324	.163	1.527	.134
Naturalness	*5.841*	*.411*	*3.838*	*.000*
Conscientiousness	*-4.497*	*-.316*	*-2.955*	*.005*
Agreeableness	*4.575*	*.322*	*3.006*	*.004*
Emotional stability	-2.471	-.174	-1.624	.111
Openness to experience	*3.329*	*.234*	*2.187*	*.034*

Note: Dependent variable: Intention; $N = 52$; $R = .696$; $R^2 = .484$; Adjusted $R^2 = .415$.

Table 3.17. Regression of purchase *intention* on the image factors of organic food products in France.

are *Naturalness* (β = .362, p < .01) and *Agreeableness* (β = .345, p = .01); while none of the six latent dimensions seem to significantly impact *perceived behavioral control* (p >.05). Directly, *Naturalness* (β = .411, p < .001), *Agreeableness* (β = .322, p < .005) and *Openness to experience* (β = .234, p < .05) influence positively *intention*; whereas *Conscientiousness* shows a negative influence on it (β = -.316, p = .005).

3.5. General discussion

The aim of the present study was, on the one hand, to understand the extent to which the moral dimension is important for organic food product purchasing, extending the TPB to *moral norms* and *moral disengagement*; and, on the other hand, to appraise the influence of organic food product personality traits on the determinants of the purchase *intention*, by employing the *Prospect method* to achieve such a purpose. The study was carried out both in Italy and in France in an effort to evaluate from a cross-cultural perspective the way consumers of these two countries' each perceive organic food products' images and express their purchase *intention*.

As regards to the first objective, the results show that in both countries, in line with what was hypothesized, *moral norms* positively influence the purchase *intention*, whereas, as predicted, *moral disengagement* shows a negative beta, albeit not one which is statistically significant (p > .05). The moral dimension hence plays a fundamental role in determining intentions: the act of purchasing organic food products, in fact, implies a series of ethical considerations regarding the consequences that the consumption of these products may have on one person's and his/her loved ones' health, in particular, and on the environment, in general. Consumers who feel morally involved show, therefore, a higher purchase intention for this kind of product. *Moral disengagement* is statistically of little weight, probably because of the "social desirability" bias related to the data collection method (Fisher

1993), a phenomenon that could distort its measurement. As maintained by Fisher (1993: 303), "the basic human tendency to present oneself in the best possible light can significantly distort the information gained from self-reports. Respondents are often unwilling or unable to report accurately on sensitive topics", such as indulging in specific forms of behavior that are in contrast with one's moral codes. As a consequence, data could be systematically distorted in relation to what is perceived as correct or socially acceptable. An additional motivation may be linked to the fact that the *moral disengagement scale* (Caprara *et al.* 1996) includes items that are too generic to be used to test the *moral disengagement* for food product consumers.

As regards to other variables which influence organic food product purchase *intention*, the most relevant predictor in Italy appeared to be *perceived behavioral control* – also significant in France immediately after *moral norms* – which, in both countries, is influenced by the control beliefs pertaining to the effects on health, as well as to the selection and to the availability of organic food products. This result mirrors that which has already emerged in other studies which consider crucial, in the purchase decision, certain values, such as healthiness (Chinnici, D'Amico and Pecorino 2002; Lockie *et al.* 2002; Makatouni 2002; Padel and Foster 2005), as well as assortment and availability (Lea and Worsley 2005). The price, considered by some other authors as an obstacle to purchase (Lea and Worsley 2005; Padel and Foster 2005; Zanoli and Naspetti 2002), was shown to be less relevant.

Furthermore, by examining the items of the *moral disengagement scale* that have the lowest mean scores, it is possible to identify which topics organic food product consumers in Italy and France declare to be less morally disengaged. In particular, organic food product consumers in both countries showed that they indeed take into consideration ethical topics related to the environment (disagreeing with statements like: "It doesn't make sense for an individual to care about environmental damage, since harmful effects are produced at a collective level"; or "Companies that produce things of public interest shouldn't be prosecuted if they pollute the environment"), in line with the importance attached to environmental considerations in the decision to purchase organic food products (in line with Honkanen, Verplanken and Olsen 2006; and Lockie *et al.* 2002).

As for the second research objective, from the analysis of the latent image dimensions, we can infer that *Extroversion* is the main personality trait of organic food products in Italy, saturated by Big Five attributes such as *jolly* and *energetic*, but also by non-Big Five attributes such as *tasty* and *good*. Therefore, organic food product consumers in Italy seem to seek, mainly, taste, as well as enjoyment while eating (Zanoli and Naspetti 2002). *Naturalness*, saturated by attributes such as *genuine*, *safe*, *healthy*, and *natural*, turned out to be the main image dimension of organic food products in France. Dimensions like *Extroversion* in Italy and *Naturalness* in France are also those influencing most the purchase *intention* of organic food products in these two countries. A possible explanation for this can be found in Hofstede's (1984) theory, which identifies five main dimensions that are significant for cross-country comparisons: *Power distance*, *Individualism/Collectivism*, *Masculinity/Femininity*, *Uncertainty avoidance*, and *Long-term/short term orientation*. Even though differences between France and Italy are not of great magnitude, due to geographical proximity and common cultural roots, France has a higher level of *Uncertainty avoidance*, which is defined as "the degree to which the members of a society feel uncomfortable with uncertainty and ambiguity" (Hofstede 1984: 83). This tendency to minimize risk can explain the greater importance attached in France to genuineness, safety, healthiness, reliability and naturalness in the organic food product image. In Italy, elements linked to more selfish themes, such as taste and quality, seem to prevail. Organic food products are furthermore perceived as jolly and energetic. Italy, in fact, as a country, presents a slightly higher level of *Individualism*, defined by Hofstede (1984: 83) as the "social framework in society wherein individuals are supposed to take care of themselves and their immediate families only".

3.6. Implications and future research

This research could have interesting implications for organic food product marketers. In particular, from the analysis of the salient beliefs of *perceived behavioral control*, i.e. the main determinant of the purchase *intention* in Italy, it emerges that *availability* and *variety* of organic food products are major aspects on which to focus in order to stimulate demand (Fotopoulos and Chryssochoidis 2000). In this perspective, a fundamental role is played by the *distribution* of these products, usually sold in specialized stores and, in recent years, also in supermarkets. If on the one hand the choice to distribute them in supermarkets can increase the potential demand, on the other, consumers have an ambivalent attitude towards them. Trust is, in fact, a fundamental factor in the decision of where to buy such products, and consumers trust both supermarkets and big companies less. However, despite the fact that specialized stores are considered more trustworthy, the many engagements of everyday life usually lead consumers to renounce to their values when faced with the greater selection and convenience of products in a supermarket (Padel and Foster 2005).

Price can play a fundamental role in explaining the difference between intentions and actual purchase behavior, as suggested by several studies (Lea and Worsley 2005; Padel and Foster 2005), even though price does not emerge here among the more salient beliefs of *perceived behavioral control*. Indeed, according to some authors (Lea and Worsley 2005; Zanoli and Naspetti 2002), a reduction of price would be appropriate in order to promote a greater diffusion of organic food products, since their high costs for consumers can be a purchase obstacle for those who would like to buy them but lack adequate financial resources. However, not all studies agree with this statement (Chinnici, D'Amico and Pecorino 2002; Hill and Lynchehaun 2002; Padel and Foster 2005). Hill and Lynchehaun (2002), for example, suggest that prices can encourage consumers to shape their opinions on the quality and taste of organic food products, on the grounds that higher prices can be related to a higher quality and even a difference in taste.

The main findings of this research, related to the latent dimensions of the organic food product image, provide indications as regards possible improvements in *communication* plans too. Companies in this field, in Italy, should in fact broadcast a message based on the concepts of the *extroversion* and *quality* of the organic food products by using marker attributes with a higher level of saturation, like *tasty*, *good*, *jolly* and *energetic*. In France, instead, communication should be based on the *safeness*, *healthiness* and *naturalness* of these products via a message capable of inspiring confidence in consumers, by making use of adjectives such as *genuine*, *safe*, *healthy*, *reliable*, and *natural* – due to the higher level of uncertainty avoidance characteristic of the French (Hofstede 1984). In this perspective, a significant role could also be played by packaging and labeling (Fotopoulos and Chryssochoidis 2000), which allow product identification and appreciation. Some studies have highlighted how reliable labels are an important factor in organic food product purchase (Fotopoulos and Chryssochoidis 2000).

This study can be used as a starting point for future studies, in that it stresses the opportunity for developing various research paths: firstly, an analysis of the relationship between intention and purchase behavior of organic food products, already attempted by some authors (Padel and Foster 2005; Tarkiainen and Sundqvist 2005). Secondly, an investigation of the methods of measuring indirectly *moral disengagement* or of substituting the *Moral disengagement scale* items with less invasive ways of questioning. Finally, from the conducted analysis it appears appropriate that future research should be carried out on a larger sample and with methods different from the one adopted in this survey, such as direct observation, in order to permit a deeper understanding of organic food product purchase not only as a result of rational decision-making, but also as a study of consumers' emotions, which are influential in purchase behaviors and in the creation of consumers' experiences.

3.7. Conclusions

In sum, the present research has demonstrated how the moral dimension plays a fundamental role in organic food product purchase, since, in both countries under analysis, *moral norms* turned out to be a significant predictor of the purchase intention, while *moral disengagement* showed a negative correlation with the intention, even though it was not statistically significant. This study has furthermore provided indications regarding the latent dimensions of the organic food product image which can influence purchase *intention* the most: i.e., *Extroversion* in Italy and *Naturalness* in France. Hence, it seems necessary that, in defining their strategies to increase sales and to support the high growth rates that the market has experienced in recent years, marketers should bear in mind the higher sensibility of organic food product consumers in relation to ethical topics.

Gianluigi Guido, M. Irene Prete and Giovanni Pino

4. The impact of ethical self-identity and safety concerns on attitudes and purchasing intentions of organic food products

4.1. Introduction

Organic food is broadly perceived as a natural alternative to conventionally produced food (Williams and Hammitt 2000). Whereas modern farming relies on synthetic chemicals, antibiotics, growth hormones, etc., the organic production system makes use of animal waste, biological fertilizers, and natural pesticides (Magkos, Arvaniti and Zampela 2006). As a consequence, healthiness and environmental friendliness are common perceptions associated with organic farming: worldwide, the majority of consumers believe that organic products are safer than conventional food and less damaging to the environment (Ureña, Benabéu and Olmeda 2008).

In the past few decades, public interest in the origin and the nutritional properties of food, as well as concerns about food-related risks, have sharply increased: the outbreak of a series of crises, like the dioxin scare or the "mad-cow" disease, has lowered indeed consumers' trust in the food industry (De Jonge *et al.* 2004). A Eurobarometer survey (the so-called annual opinion pool commissioned by the European Union on topics of public interest) carried out in 2005 in twenty-five EU member states revealed that approximately a quarter of the respondents were frightened of food contamination incidents due to polluting agents or new strains of virus. Previous research, however, demonstrated that individuals differ in their perception of food safety hazards (cf. Williams and Hammitt 2001): compared to conventional food purchasers, organic food buyers have been found to evaluate a higher annual rate of mortality caused by agro-chemical residues and to believe that organic food is safer than conventionally grown produce. In addition, it has been shown that organic food buyers are more likely to recycle household rubbish and to purchase eco-labeled cleaners (Williams and Hammitt 2000). Various studies have associated organic food purchasing to a consumption pattern centered around the preservation of the natural ecological balance and biodiversity (e.g., Magnusson *et al.* 2003; Makatouni 2002). Such a behavioral pattern, originally known as "green consumerism", has subsequently broadened its meaning, and nowadays "green" – or, in general, "ethical" – consumers are likely not only to buy eco-labeled products, but also to get involved in boycott initiatives aimed at attaining social relevant objectives (see Freestone and McGoldrick 2008).

The relationship between consumers' sensitivity to ethical issues and willingness to buy products like organic food and fair trade grocery has been investigated by a series of quantitative studies in Europe, most of which have been carried out in the UK (e.g., Arvola *et al.* 2008; Shaw and Shiu 2002a; Sparks and Shepherd 1992). British consumers' beliefs have also been surveyed by qualitative research and often compared to those of people living in different nations – for example, Germany (see Baker, Thompson and Engelken 2004) and other UE countries (see Naspetti and Zanoli 2006) – with the aim of depicting the linkages existing among the intention to purchase of organic foods, and psychological dimensions like worries about food-related risks and ethical awareness. In Italy, quantitative research in this area is relatively scarce and has not attempted to compare the drivers of British and Italian consumers' preference for organic food.

To fill this gap, the present study addresses the topic of organic food consumption with a twofold purpose: (*i*) to examine the effect exerted by concerns relating to food safety and ethical issues on individuals' attitudes and purchasing intentions of organic food products; and (*ii*) to extract further information about the determinants of organic food purchase behavior by observing the relationships among the aforementioned dimensions from an intercultural perspective.

4.2. Antecedents of organic food purchase intentions

To date, food choice behavior has been extensively depicted through the *Theory of planned behavior* (hereafter TPB, Ajzen 1991). This conceptual framework postulates that *Attitude* towards a certain object, together with *Subjective norm* and *Perceived behavioral control*, are the strongest predictors of a person's *Intention* to engage in a particular behavior. *Intention* represents an individual's motivation for enacting a given behavior; *attitude* is defined as a "propensity to evaluate a certain entity with some degree of favorability or unfavorability" (Eagly and Chaiken 2007: 583) and results from a person's beliefs with regard to the specific object; *subjective norm* refers to the pressure exerted on the individual by other persons considered important, who may approve or disapprove the individual's decision to perform that particular behavior; whereas *perceived behavioral control* reflects the extent to which an individual believes that the behavior in question is under control (Ajzen 1991).

As noted above, "absence of pesticide residues" and "pureness" are widespread beliefs associated with organic foods: a number of qualitative researches have indeed proved that a positive *attitude* towards such products is mainly shaped by consumers' *food safety concerns* (e.g., Baker, Thompson and Engelken 2004; Von Alvensleben 2001). Few studies, however, have tested this link on a quantitative ground (e.g., Magnusson *et al.* 2003; Saba and Messina 2003) and little is known about *how* concerns about the safety of food products affect organic food purchasing decision.

Past research also found support for the inclusion of additional determinants in the standard version of the TPB (Shaw, Shiu and Clarke 2000): inasmuch as it has been acknowledged that a social conduct, in certain behavioral contexts (like shoplifting, lying, cheating, etc.), stems from moral considerations and from people's self-perception, measures of *Moral norms* and *Self-identity* have been integrated in the traditional model configuration. *Moral norms* refer to a system of "internalized ethical rules" (Shaw, Shiu and Clarke 2000: 882), and reflect one's "responsibility to perform, or refuse to perform a certain behavior" (Ajzen 1991: 199). This measure has been shown to have a positive effect on organic food purchase *intention* (Arvola *et al.* 2008; and also Chapter 2, *above*) and fair trade grocery purchase *intention* as well (Shaw, Shiu and Clarke 2000) – hence contributing to enhance the explanatory power of the original TPB (Ajzen 1991). *Self-identity* has been defined as a set of "enduring characteristics that people ascribe to themselves" (Sparks and Guthrie 1998: 1394) and reflects the influence exerted by social membership on a person's behavior. Measures like "role-identity", referring to a person's position in a social context, and "social-identity", which expresses people's tendency to identify themselves with their social groups, can be considered as peculiar forms of *self-identity* (Sparks and Guthrie 1998). A number of studies have demonstrated the existence of a predictive effect of this measure mainly with respect to consumers' *intention* to purchase organic food (e.g., Sparks and Shepherd 1992) and fair trade products (Shaw, Shiu and Clarke 2000). Sparks and Guthrie (1998), however, highlighted the importance of assessing the existence of a linkage between *self-identity* and consumers' *attitude*, too. As regards the purchase of organic food, such a relationship was proved in Chapter 2 (*above*) by regressing *attitude* on both *moral norms* and *self-identity*, while, with respect to fair trade products, an equivalent result was obtained by Shaw and Shiu (2002a).

Shaw, Shiu and Clarke (2000) also drew attention to the link between *moral norms* and *self-identity* arguing that "ethical consumers may make ethical consumption choices because ethical issues have become an important part of their self-identity" (2000: 882). Similarly, Sparks and Shepherd (1992) pointed out that, in certain cases (e.g., blood donation, and "green" consumption), individuals' *self-identity* may entail a perception of ethical concerns. Such reflections inspired a recent study by Michaelidou and Hassan (2008) who proposed a measure of the consonance between a consumer's identity and the image of the "ethical consumer". By aggregating the dimensions of *moral norms* and *self-identity*, Michaelidou and Hassan (2008) developed a meta-construct referred to as *Ethical self-identity*, which was shown to affect both consumers' *attitude* towards organic food and the *intention* to buy such products.

It should be added that the strength of the link between ethical considerations, *attitude* towards organic food, and the consequent purchase *intention* – as well as that between *food safety concerns*, *attitude*, and purchase *intention* – is likely to change from country to country. National cultures seem to differ, in fact, in their tolerance for risky events or in their level of integration and involvement with societal issues. Both aspects have been examined in Hofstede's (1983, 1984) work on "cultural distance" by means of two indicators named *Uncertainty avoidance* and *Individualism*. Together with a small number of other cultural dimension indexes, ranging from zero to 100, these indicators enabled Hofstede to depict the cross-national profiles of more than fifty different countries.

The *Uncertainty Avoidance Index* (or UAI), in particular, expresses a group's rejection of indefinite situations and a need for explicit rules, while the *Individualism Index* (or IDV) measures the extent to which a group's members care about social cohesion and are willing to interact with each other (Hofstede 1983). *Uncertainty avoidance* has been demonstrated to be inversely related to a group's acceptance of new products (Yeniyurt and Townsend 2003). For example, the higher the level of risk aversion of a certain nation, the lower may be the diffusion of products like cellular phones and personal computers. In the field of food consumption, Van Dijk *et al.* (2008) argued that, depending on the level of *Uncertainty avoidance* characterizing the different groups, international policies relative to food hazards may be perceived as more or less effective by consumers living in different countries. Wherein low levels of *Uncertainty avoidance* have been detected, consumers are likely to trust regulatory bodies and feel confident with their management practices. Conversely, groups holding high levels of *Uncertainty avoidance* are likely to ask for more security and stricter regulations. These authors, therefore, highlighted the need to fine-tune international food safety policies to the cultural features identifying the different nations.

As regards *Individualism* and its relationship with consumption behavior, it should be pointed out that such a dimension appears to be closely connected with ethics. It could be argued, in fact, that an inverse relation exists between *Individualism* and ethics: the more people strive for their private interests, the more likely they will be to transcend the needs of other persons and the welfare of nature (Springer *et al.* 2005). Such association, for example, has been demonstrated by Gerganov *et al.* (1996) for Bulgarian citizens: in their study, subjects holding an individualistic orientation proved, in effect, to attach limited importance to ethical concerns and interpersonal ties.

4.3. Purpose of the study and methodological framework

The present study is conceived as a replication and extension in Italy of the survey conducted in the UK by Michaelidou and Hassan (2008). More specifically, it addresses two primary determinants of organic food purchasing that were taken into consideration by the previous study, namely consumers' concerns about food safety issues (i.e., *food safety concerns*) and their degree of identification with the ethical consumers' image (i.e., *ethical self-identity*).

To analyze how these dimensions affect individuals' *attitude* towards organic food and, possibly, purchase *intention*, a *mediation model* was adopted. Shedding light on the mechanism through which a particular effect occurs, mediation analysis allows one to gain a deeper insight into the nature of several decision-making processes (e.g., Eartmans *et al.* 2005; MacKinnon, Fairchild and Fritz 2007). This methodological approach postulates that the link between a certain antecedent and an outcome of interest is influenced by a third variable acting as mediator. According to this model, the mediator is posited to be responsible for the transmission of the effect that an alleged independent variable has on a dependent variable (Baron and Kenny 1986). Such an effect, referred to as *total effect*, results from the amount of the *indirect effect* exerted by the independent variable on the dependent variable through the mediator, plus the amount of the residual predictive effect directly exerted by the independent variable on the dependent variable, referred to as *direct effect* (MacKinnon, Fairchild and Fritz 2007).

In this study, *attitude* towards organic food is hypothesized to mediate the effects shaped by *food safety concerns* and, alternatively, by *ethical self-identity* on purchase *intention*. That is to say, that also in Italy, as well as in UK, *attitude* is hypothesized to be a proximal determinant of purchase *intention*, and to be affected, in turn, by *food safety concerns* and *ethical self-identity*. Evidence for a predictive effect of consumers' dispositions towards organic food products on buying *intention* has been provided by several studies (e.g., Honkanen, Verplanken and Olsen 2006; Magnusson *et al.* 2003). As a consequence, it is here supposed that:

H1: Organic food purchasing intention is directly predicted by consumers' attitude.

As previously observed, the TPB model (Ajzen 1991) postulates that an individual's *attitude* towards a particular object stems from his/her relevant beliefs associated with the *attitude* object. Since consumers' concerns about food safety risks have been proved to be a major motive for buying organic food (e.g., Von Alvensleben 2001; Williams and Hammitt 2000), the validity of the following statement is proposed:

H2: Organic food purchasing intention is indirectly predicted by food safety concerns, with attitude acting as mediator.

Finally, insofar as *moral norms* have been shown to indirectly affect organic food purchase *intention* through consumers' *attitude* (Arvola *et al.* 2008), it could be supposed that an analogous pathway can lead from *ethical self-identity* to organic food purchase *intention*. For this reason, it is here hypothesized that:

H3: Organic food purchasing intention is indirectly predicted by ethical self-identity, with attitude acting as mediator.

4.4. Methodology

This study was directed at replicating and extending Michaelidou and Hassan's (2008) study from UK to Italy, thus assessing possible intercultural differences between these two European countries. To collect information, a self-explanatory questionnaire was drawn up from the original survey instrument developed by the above-mentioned authors. Data were collected during a period in which no food contamination outbreaks were reported by the media.

4.4.1. Sampling

Over the last decade, there was a ten-fold increase in the number of primary school canteens supplied with organic produce in Italy (Bertino 2006). It seems reasonable to predict that this trend will soon boost the offering of organic food into university campuses, too. Hence, also to assess the potential of such a market-segment, the present survey was administered to a sample of 261 undergraduate students randomly selected from different faculties of a medium-sized Italian university. The majority of the students were female (70%) and their average age was 22 years (SD = 2.87). Respondents were approached in quiet locations, such as libraries and lecture halls. They were introduced to the purpose of the survey by a researcher and assisted during the completion of the questionnaire, which lasted approximately fifteen minutes.

4.4.2. Measures

The questionnaire, originally written in English, was translated into Italian by two lecturers at the same Italian university. One of them was a bilingual born in an English-speaking country, whereas the second one – whose mother tongue was Italian – was a full professor of English. Subsequently, two different bilingual lecturers – whose mother tongue was English – back-translated the questionnaire into the original language. By comparing this version to the original one, mistranslations from English to Italian were checked and accurately corrected in order to capture the exact meaning of the questions included in the original survey instrument.

Prefaced with a brief definition of "organic food products", the questionnaire was structured into three sections, measuring: respondents' *intention* to buy organic food, the *attitude* towards such products, and two other constructs – namely, *food safety concerns* and *ethical self-identity* – alleged to be both indirect predictors of the buying *intention*. Socio-demographic data, like gender and age, were gathered, too.

Purchase *intention* was measured using two items, taking the form: "I intend to purchase organic food produce within the next fortnight" and "I want to purchase organic food produce within the next fortnight". Answers were collected on a seven-point (0 to 6) categorical scale, ranging from "Not at all" to "Definitely", and from "Definitely do not" to "Definitely", respectively. *Attitude* towards organic food was assessed by means of two items and measured on a seven-point (-3 to +3) semantic differential scale, ranging from "Unfavorable" to "Favorable", and from "Negative" to "Positive", respectively. Both scales of *attitude* and buying *intention* were built following Ajzen's (1991) recommendations.

Subjects' concerns about food safety issues and identification with "ethical consumers" were surveyed by using two sets of items, previously tested by Roddy, Cowan and Hutchinson. (1996) and by Shaw, Shiu and Clarke (2000), respectively. Both constructs were appraised through a seven-point Likert scale ranging from "Strongly disagree" (corresponding to -3) to "Strongly agree" (corresponding to +3). Specifically, *food safety concerns* were measured on a three-item scale by referring to all those agents (pesticides, additives, etc.) which may represent a potential threat to human health (e.g., "Nowadays most food contain residues from chemical sprays and fertilizers"). Analogously, *ethical self-identity* was measured by a synthetic scale consisting of two items ("I think of myself as someone who is concerned about ethical issues", and "I think of myself as an ethical consumer") that was drawn from prior studies about fair trade products

(Shaw, Shiu and Clarke 2000; Shaw and Shiu 2002a) and organic food products (Sparks and Shepherd 1992).

While the scale tested by Shaw, Shiu and Clarke (2000) also included items asking respondents to rate the importance of a series of specific ethical issues (such as "human rights", "animal welfare", etc.), the one developed by Michaelidou and Hassan (2008) consisted of more general statements enabling respondents to freely interpret the concept of "ethical consumerism" according to their feelings or their familiarity with the term. Although this scale captured limited information about consumers' beliefs, the choice to use a limited number of propositions that did not refer to a particular ethical issue appears coherent with Michaelidou and Hassan's (2008) aim of examining the interrelationships existing among individuals' *intention* to buy organic food and the *direct* and the *indirect* antecedents of such a psychological dimension.

4.5. Results

Data analysis was carried out in three steps: firstly, the internal reliability of the scales was checked; secondly, a descriptive analysis was performed; finally, research hypotheses were tested. The following sections provide details about the survey results.

4.5.1. Reliability analysis

The internal consistency of the scales was checked by calculating Cronbach's α coefficients. Table 4.1 (*below*) shows that the average inter-item correlation was quite high for most of the constructs (Cronbach's $\alpha \geq .80$). Only the coefficient pertaining to *food safety concerns* was just below the threshold of acceptability ($\alpha = .70$). However, taking into account that the scale employed consisted of only three items,

Variables	*Cronbach's α*	*Explained variance*	*Eigenvalues*
INT	.968	94.332	1.886
ATT	.939	96.993	1.940
FSC	.657	60.132	1.804
ESI	.880	83.668	1.673

Note: $N = 261$; INT = Intention; ATT = Attitude; FSC = Food safety concerns; ESI = Ethical self-identity.

Table 4.1. Internal reliability analysis: result summary.

such a result can be considered satisfactory (cf. Field 2005). The moderate level of internal consistency cannot be ascribed to misinterpretations occurred during the phase of translation of the original survey instrument: a similar pattern of results was obtained, in fact, by Michaelidou and Hassan (2008).

Next, separate *principal component analyses* were performed to assess the extent to which the complete set of items used for each scale could be reduced to a single dimension. All of the constructs were found to be one-dimensional and the highest percentages of explained variance were observed for *intention* and *attitude* (96% and 94%, respectively).

4.5.2. Descriptive analysis

Means and standard deviations of each measure were computed to summarize the collected data (Table 4.2). Results revealed that

Variables	*M*	*SE*	*SD*
INT	3.350	.112	1.811
ATT	2.017	.072	1.177
FSC	1.632	.066	.954
ESI	1.070	.059	1.080

Note: $N = 261$; INT = Intention; ATT = Attitude; FSC = Food safety concerns; ESI = Ethical self-identity.

Table 4.2. Descriptive statistics.

subjects' *attitude* towards organic food products was eminently positive: on a scale ranging from -3 to +3, a mean value equal to 2.017 (SD = 1.177) was computed. On a sample of 261 subjects, 222 (85%) expressed a favorable attitude towards these products. However, this favorable disposition was not reflected in a strong purchase *intention*: out of the 261 respondents, 152 (58%) rated their average *intention* below the scale midpoint of 3. Moreover, the selected sample appeared to be quite concerned about food safety (79%) and sensitive to the ethical consumerism issue (67%).

4.5.3. Hypotheses testing

Pearson's correlation coefficients (r) were computed to assess the presence of linear associations among the examined variables. Results, presented in Table 4.3, suggest that all the study constructs were significant ($p < .01$) and positively correlated. More specifically, the weakest correlation was observed between *food safety concerns* and *ethical self-identity* ($r = .366$); the strongest association was found for *attitude* towards organic food and buying *intention* ($r = .658$). The existence of a significant relationship between respondents' *attitude* towards organic food products and the *intention* to purchase these products was established by means of a standard linear regression model (Table 4.4, *below*). In particular, *attitude* was entered in the regression equation as the independent variable, whereas *intention* was assumed to be the dependent one. The model yielded an adequate fit ($R^2 = .433$) and returned *attitude* as a significant predictor of purchase *intention* ($\beta = .658$; $p < .005$). Hypothesis H1 was, therefore, confirmed.

	INT	ATT	FSC	ESI
INT	1.000	-	-	-
ATT	.658*	1.000	-	-
FSC	.376*	.497*	1.000	-
ESI	.409*	.497*	.366*	1.000

Note: $N = 261$; INT = Intention; ATT = Attitude; FSC = Food safety concerns; ESI = Ethical self-identity; * = Significant at the .01 level.

Table 4.3. Pearson's correlation coefficients.

Variables	*B*	*Beta*	*t*	*p*
Constant	1.308		7.780	.001
ATT	*-.012*	*.658*	*14.057*	*.001*

Note: Dependent variable: Intention; N = 261; R = .658; R^2 = .443; Adjusted R^2 = .431; ATT = Attitude.

Table 4.4. Regression of purchase *intention* on *attitude* towards organic food products.

To ascertain the mediating role of *attitude* towards organic food in the relationship between *food safety concerns* (alleged to be the predictor variable) and purchase *intention* (alleged to be the outcome variable), a series of tests was run (see Table 4.5). Firstly, purchase *intention* was regressed on *food safety concerns* to evaluate the overall impact (total effect) of the predictor on the dependent variable (*Step 1*: R^2 = .141, β = .376, $p < .005$); next, by regressing *attitude* on *food safety concerns* it was found evidence of a significant association between the predictor and the mediator (*Step 2*: R^2 = .247, β = .497, $p < .005$); finally, *food safety concerns* and *attitude* were simultaneously

Steps	*Variables*		*B*	*Beta*	*t*	*p*
1.		Constant	2.186		10.583	.001
	Independent:	*FSC*	*.713*	*.376*	*6.524*	*.001*
	Dependent:	INT				
2.		Constant	1.016		8.086	.001
	Independent:	*FSC*	*.613*	*.497*	*9.221*	*.001*
	Dependent:	ATT				
3.		Constant	1.208		6.433	.001
	Independent:	FSC	.123	.065	1.199	.232
		ATT	*.963*	*.626*	*11.611*	*.001*
	Dependent:	INT				

Note: N = 261; $R^2_{(1)}$ = .141; Adjusted $R^2_{(1)}$ = .138; $R^2_{(2)}$ = .247; Adjusted $R^2_{(2)}$ = .244; $R^2_{(3)}$ = .436; Adjusted $R^2_{(3)}$ = .432; INT = Intention; ATT = Attitude; FSC = Food safety concerns.

Table 4.5. *Attitude* as mediator of the relationship between *food safety concerns* and purchase *intention*: results of mediation test.

entered in a multiple regression equation as predictors of *intention* (*Step 3*: $R^2 = .436$). As in this test the relationship between *food safety concerns* and purchase *intention* proved to be no longer significant ($\beta = .065$, $p > .05$), it can be stated that a *complete* mediation process occurred. That is to say, the effect of *food safety concerns* on purchase *intention* was entirely transmitted by the *attitude* towards organic food.

Moreover, it was possible to decompose the *total effect* exerted by the predictor variable on the dependent variable as the sum of a *direct effect* and an *indirect effect*. The *direct effect* was given by the beta weight obtained in the third equation with respect to the predictor variable ($\beta = .065$, $p > .05$). The *indirect effect* was computed by multiplying the beta weight determined by means of the second regression equation ($\beta = .497$, $p < .005$) by the beta weight calculated in the third equation and referring to the mediator variable ($\beta = .626$, $p < .005$). Such a computation can be summarized as follows: *total effect* = [*direct effect* + (*indirect effect*)] = [.065 + (.497 × .626)] = .376.

To check the significance of the *indirect effect*, the null hypothesis that it equals zero was held. Next, a critical ratio (*z-value*) was calculated by dividing the *indirect effect* by its standard error: denoted as a and s_a the beta weight determined in *Step 2* and its standard error, respectively (see Table 4.5, *above*); and denoted as b and s_b the beta weight determined in *Step 3* with respect to the mediator variable and its standard error, respectively (see Table 4.5, *above*), the critical ratio was computed as indicated in (1). As it was greater than the critical value from the standard normal distribution ($z = 1.96$), the null hypothesis was rejected, thus confirming Hypothesis H2.

$$z\text{-value} = (\mathrm{a} \times \mathrm{b}) / \sqrt{b^2 s_a^2 + a^2 s_a^2 + s_a^2 s_b^2} = 3.8 \qquad (1)$$

Following the procedure presented above, the interrelationships among *ethical self-identity*, *attitude* towards organic food, and purchase *intention* were examined. To test Hypothesis H3, a mediation model was built assuming *ethical self-identity* to be an antecedent of purchase *intention*, and *attitude* towards organic food to be a mediator variable (Table 4.6, *below*). The linkage between *ethical self-identity* was appraised at first, as a prerequisite for establishing a mediation effect (*Step 1*: $R^2 = .168$, $\beta = .409$, $p < .005$). Thereafter, *ethical self-identity* was demonstrated to have an effect on *attitude* (*Step 2*: $R^2 = .247$, $\beta = .497$, $p < .005$). And, finally, purchase *intention* was regressed on both the predictor and the mediator variables (*Step 3*: $R^2 = .442$). Results show that the mediator reduced the effect of the predictor on the dependent variable; the effect the predictor exerted on the dependent variable, however, remained significant ($\beta = .109$, $p < .005$). In this case, therefore, *partial mediation* occurred, that is, purchase *intention* was mostly predicted by *attitude* and for a small – but non negligible – measure by *ethical self-identity*, too. As a consequence, Hypothesis H3 was accepted and the overall predictive effect of *ethical self-identity* on purchase *intention* was decomposed as indicated in the following equation: *total effect* = [(*direct effect* + *indirect effect*)] = [.109 + (.497 × .604)] = .409.

Steps	*Variables*		*B*	*Beta*	*t*	*p*
1.		Constant	2.616		18.105	.001
	Independent:	*ESI*	*.686*	*.409*	*7.220*	*.001*
	Dependent:	INT				
2.		Constant	1.437		16.100	.001
	Independent:	*ESI*	*.542*	*.497*	*9.227*	*.001*
	Dependent:	ATT				
3.		Constant	1.281		7.638	.001
	Independent:	*ESI*	*.183*	*.109*	*2.035*	*.001*
		ATT	*.929*	*.604*	*11.257*	*.001*
	Dependent:	INT				

Note: $N = 261$; $R^2_{(1)} = .168$; Adjusted $R^2_{(1)} = .164$; $R^2_{(2)} = .247$; Adjusted $R^2_{(2)} = .244$; $R^2_{(3)} = .442$; Adjusted $R^2_{(3)} = .437$; INT = Intention; ATT = Attitude; ESI = Ethical self-identity.

Table 4.6. *Attitude* as mediator of the relationship between *ethical self-identity* and purchase *intention*: results of mediation test.

4.6. Discussion and implications

The overall findings of this study were in line with those published by Michaelidou and Hassan (2008), thus allowing one to generalize results from UK to another European context, namely Italy. As expected, *attitude* played a critical role in explaining individuals' intentions to purchase organic foods. It should be noted, however, that, as reported by Michaelidou and Hassan (2008) and by previous research (e.g., Magnusson *et al.* 2001), also in this study a substantial proportion (29%) of subjects who expressed a positive disposition towards organic food declared to be unwilling to buy them.

Respondents' intentions to purchase organic food proved to be affected by concerns related to food safety issues and by ethical motivations. As in the case of the original study, both the predictors were demonstrated to influence *intention* via *attitude* towards organic food. However, contrary to the results by Michaelidou and Hassan (2008), in this survey a weak, but significant, relation between *food safety concerns* and organic food purchase *intention* was detected. Such a discrepancy can be attributed, at a cross-cultural level, to Hofstede's (2009) *Uncertainty Avoidance Index* (UAI, see Section 4.2, *above*) which expresses a group's reluctance to get involved in unknown situations: compared to that determined for the UK (UAI_{UK} = 35), such an indicator reaches for the Italian society a far higher value (UAI_{IT} = 75) (see Hofstede 2009). Noteworthy differences in terms of food-related risk perception have also been detected between the two countries through the Eurobarometer (2006) survey: allergic reactions and chemical residues were shown to frighten 59% and 86% of the Italian respondents, respectively. The same indexes were found to be considerably lower in the UK: levels of 29% and 65%, respectively, were ascertained, in fact, in these cases.

Besides, it should be added that a cross-cultural research by Bredhal (1999) reveals that Italian food choice behavior, more than that in the UK, is heavily influenced by concerns about food scares or food scandals. In this study, the *Security* value was indeed demonstrated to be a major determinant of Italian consumers' rejection of genetically modified food. As a result, it could be stated that properties like naturalness and healthiness should be opportunely highlighted to appeal to the Italian demand for organic foods, whereas, from this perspective, more efforts seem to be requested to marketers and policy-makers to stimulate an analogous interest in these products among British consumers.

The results of the present study also suggest that the relationship between *food safety concerns* and purchase *intention* is *fully* mediated by *attitude* towards organic food. Such a result implies that food-related risks are more likely to generate a favorable disposition towards the examined products, rather than immediately influence purchase intention. Indeed, Naspetti and Zanoli (2006) refer to the safety attributes of the products in question as "sleeping criteria", which act as primary buying motivations mainly when food contamination accidents and food scandals occur. Thus, making salient the health properties of organic food will probably have the effect of corroborating the widespread positive attitude towards organic farming and its products which, in turn, is supposed to enhance consumers' buying intentions.

Innovative forms of distribution like Internet sites, enabling users to purchase food products from virtual grocery stores, as well as direct sale markets, wherein small and medium-size producers can regularly meet consumers (so-called *farmers' markets*), seem suitable tools for this task: by dropping the price of organic food, these channels can generate (or enhance) a favorable disposition towards organic products. Direct sales, in particular, by enabling purchasers to personally check the naturalness of the organic production system, are expected to match the needs for assurances about the origin of the products and strengthen, in such a way, consumers' purchasing intentions (Naspetti and Zanoli 2006). Similar conclusions could be drawn with respect to school canteens dispensing organic foods: satisfying families' needs for nourishing their children in a safe and healthy manner, organic canteens can be supposed to act as a promotional lever which may positively affect consumers' attitudes towards organic food and exert an indirect effect on their purchasing intentions.

Furthermore, this service could suit the sense of moral obligation that parents commonly perceive towards future generations. The present study – as well as that by Michaelidou and Hassan (2008) – has demonstrated, in fact, that ethical considerations are another major determinant of organic food consumption. As shown in the original study, the interaction between *ethical self-identity* and buying intentions proved to be partially mediated by attitudes towards organic food in this study, too. That is to say, *ethical self-identity* fed into purchase intentions both directly and indirectly. Slightly different beta weights relating to these paths, however, were determined in the two surveys. Compared to those assessed in the present study (see Table 4.6, *above*), the coefficients computed by Michaelidou and Hassan (2008) proved to be slightly lower ($\beta = .23$, $p < .01$, for the path from *ethical self-identity* to purchase intention; and $\beta = .31$, $p < .01$, for the path from *ethical self-identity* to attitude towards organic food). Such a divergence was probably due to cultural factors. According to Hofstede (2009), a moderate difference exists between Italy and the UK in terms of *Individualism*: on a scale ranging from zero to 100, an *Individualism Index* (ID, see Section 4.2, *above*) equal to 76 has been ascertained for the Italian cultural context, whereas a level of 89 has been established for the UK (see Hofstede 2009). Since the groups focusing on principles like autonomy and personal achievement are likely to disregard the importance of values like helpfulness, solidarity, and care for the environment (Springer *et al.* 2005), it could be argued that the higher level of *individualism* which characterizes the British culture may have had the effect of weakening the relationships among *ethical self-identity*, attitudes towards organic food, and purchase intentions.

This conclusion is supported also by the findings of a qualitative study carried out by Baker, Thompson and Engelken (2004). A series of laddering interviews detected, in this case, the strong individualistic orientation of the British population as a dominant factor affecting organic food buying behavior. The cognitive representation of consumers' decision making process, obtained through the so-called hierarchical value map, was, in effect, characterized by the absence of linkages between organic food consumption and concerns relating to environment protection. Such a result leads one to argue that ethical awareness, for the British culture, can be firstly associated with the care for one's own well-being and that of his/her relatives, and secondly, with the environment protection and the animal welfare (see also Makatouni 2002).

The validity of the relationship between ethics and consumers' buying behavior has been documented by several quantitative studies. Evidence for a noteworthy association between attitudes towards organic food and ethical motives has been reported, for example, by Honkanen, Verplanken and Olsen (2006), whereas Shaw and Shiu (2002a), focusing on fair trade grocery products, and by implementing the TPB (Ajzen 1991), have modeled the relations between *self-identity*, *moral obligation*, and buying *intention*. Taken together, these results suggest that the link between organic food purchasing and ethical causes (like reforesting projects, humanitarian initiatives, and so on) should be enforced in order to stimulate the adoption of more responsible consumption patterns and enable ethical consumers to contribute to the protection of nature and to the welfare of others.

4.7. Limitations and future research

A limitation of this survey regards the bipolar three-items scale used to assess *food safety concerns*. Several studies have indeed emphasized the importance of developing more effective measures of consumers' need for safe and healthy food (e.g., Hughner *et al.* 2007; Williams and Hammitt 2001). Emotional involvement and accessibility of specific risk-related beliefs, for example, have been indicated by Williams and Hammitt (2001) as important dimensions which deserve greater attention to better understand the exact meaning of the notion of "food safety concerns".

It should be also observed that a social desirability bias may have amplified the magnitude of the predictive effect shaped by *ethical self-identity* on participants' attitudes and buying intentions: subjects tend, in fact, to conform their public image to highly regarded standards of conduct (Freestone and McGoldrick 2008). Therefore, to more properly evaluate the effect of ethics on organic food consumption, the interaction between *ethical self-identity* and actual purchasing behavior should also be investigated.

Finally, it should be recognized that the generalizability of the obtained results can be limited: this study concentrated, in fact, on a specific subgroup of the national population. Support for reliable comparability of the results found by surveying student samples or, alternatively, housewife samples was provided by seminal research (Sheth 1970). As regards the analysis of the relationships between attitudes and behaviors, positive evaluations about the employment of student samples in empirical tests have also been expressed by Yavas (1994) and Beltramini (1983). However, to evaluate more in depth the influence exerted by cultural factors on consumers' behavioral choices, future research will have to inspect a wider population group.

4.8. Conclusions

The present study has sought to gain a better understanding of the mechanisms through which concerns about food-related hazards and people's tendency to recognize themselves as ethical consumers impact on both the intention to purchase organic food and the disposition towards these products. By replicating in Italy a study conducted in the UK, the influence of cultural factors was captured, too.

Attitude towards organic food proved to play a decisive role in shaping respondents' decisions as it resulted to act as a bridge between a person's beliefs and his/her behavioral intentions. *Food safety concerns* were demonstrated to indirectly affect individuals' purchase intentions, since they exerted most of their effect on attitude towards organic food. *Ethical self-identity* was shown to influence both participants' dispositions towards the study products and the consequent purchase intentions.

These findings may hold a critical valence for producers, institutions, and consumer associations, since they may help them set appropriate communication campaigns. Firms aspiring to boost sales could target the psychological dimension that proved to directly influence purchasing intention, that is consumers' sensitivity to ethical issues. Whereas, organizations aiming at improving people's judgments about organic foods could focus on the dimension that proved to mainly impact on the attitudes towards these products, that is consumers' concerns about food hazards.

In this chapter, it has also been argued that the choice to buy organic foods seems to reflect a nation's level of tolerance for high-risk events – like food scandals or virus contaminations – and, at the same time, appears to entail an ethical meaning. Such meaning is likely to change across cultures: according to the level of *Individualism* characterizing a certain nation, people may prioritize the care for their own health and that of their families or, alternatively, may be more sensitive to issues like sustainable consumption, animal rights, or biodiversity loss. As a consequence, it could be asserted that, to value the "potential" that ethics possesses in present-day society (cf. Parkins and Craig 2006), when formulating educational campaigns, marketing initiatives, and communication polices, in general, international regulatory bodies and firms operating in different countries should not disregard the influence that cultural diversities can exert on the receptiveness of their messages.

Gianluigi Guido, Giovanni Pino and M. Irene Prete

5. Effects of attitude and personal values on the purchase intention of genetically modified food products

5.1. Introduction

Modern biotechnologies make use of living organisms to achieve medical and industrial purposes, such as the improvement of particular processes and the development of new products. Thanks to the transferring of genetic material between different species, these techniques allow the creation of new varieties of breeding plants and animals, which are commonly known as *Genetically Modified Organisms* (GMOs) (Suslow, Thomas and Bradford 2002).

While the medical and pharmaceutical applications of biotechnologies (e.g., insulin production) have met weak, or any, opposition (Bredhal, Grunert and Frewer 1998), the use of genetically modified ingredients in the food manufacture has caused an ongoing debate, especially in Europe (Rimal, Moon and Balasubramanian 2006). According to the Eurobarometer (2006) survey, 37% of European Union citizens consider themselves "fairly worried" about the accidental presence of GMOs in food or drinks. Indeed, since their early development, genetically modified crops and farm animals have been considered a threat to consumers' health, posing unknown hazards for humans, as well as for nature and the environment (Siegrist 2003). Nevertheless, between 2006 and 2007, a solid growth rate of the global area devoted to GM crops (+12%) was registered (James 2007), whereas, throughout 2007, over 150 million of hectares of transgenic vegetables varieties were grown worldwide.

By cultivating GM plants, in fact, farmers may profit from a series of agronomic benefits, such as tolerance to drought, insects, and pesticides, as well as economic advantages, like higher and more uniform crop yields (Suslow, Thomas and Bradford 2002). By contrast, from the consumers' perspective, no beneficial effect can be directly associated to GM food consumption: biotech products are not only regarded as harmful and unnecessary, but also as morally unacceptable (Haukenes 2004). The adoption of genetic technology in the agro-food sector has been, in fact, perceived as a severe alteration of the natural order of things posing the risk of unknown effects for human health and the planetary ecosystem (Frewer *et al.* 2004). Furthermore, multinationals holding patents on transgenic seeds are currently blamed for widening the gap between small and large growers, as well as that between industrialized and developing countries (Perrez 2000).

Worldwide, consumers' associations have started to invoke peoples' right to be informed about the origin of food and to choose between GMO and non-GMO products, the latter category indicating those "foods that are reasonably free of genetically engineered (GE) material" (Roff 2008: 1). For the UE member states, in particular, the European Commission has established that non-GMO (a.k.a. "GM free") food products contain less than 0.9% of transgenic material. Such a tolerance threshold, however, has been judged too large by opponents to the applications of biotechnology in the farming sector, especially with respect to organic food, which are non-GMO products by definition (Devos *et al.* 2008). Consequently, to avoid seed contaminations, several regions across the EU member states (the so-called "GM-free zones") have prohibited the cultivation of GM crops or have imposed the segregation of GM varieties from conventional and organic ones.

Recent studies have prospected that consumers' willingness to pay a premium for non-GMO product may determine the development of a new niche market (e.g., Roff 2008). At the same time, it has been argued that consumers' skepticism towards genetic technology applications will lower once the perceived benefits deriving from the adoption of such techniques will overtake the entity of the perceived risks (Gaskell *et al.* 2004). The introduction of a second generation of GM foods, endowed with functional properties, is supposed to be one of the key-factors which can enhance acceptance of biotechnology in the agro-food sector (Laure *et al.* 2004). Though genetic engineering represents a dispensable requisite in functional food production, the development of such a new market segment can be expected to have a positive impact on the growth of the GM food industry (Purchase 2005). Acceptance of GM products, however, is likely to change across countries (Springer *et al.* 2005): cultural factors impact, in fact, on consumers' perceptions of risks and benefits associated with biotechnologies and technology in general, and this may facilitate or hinder the diffusion of new food products (Frewer *et al.* 2004; Hossain and Onyango 2004; Kolodinsky, DeSisto and Labrecque 2003; Yeniyurt and Townsend 2003).

This study addresses cultural values as a determinant of the disposition towards GM food and the intention to purchase such products, and seeks to compare the behavioral patterns of consumers living in two different nations, namely Italy and Norway. An overview of the frameworks developed in literature to appraise cultural divergences among countries is presented in the following sections. Next, the essential features of consumers' dispositions towards GM food are described. Once the research purpose and the adopted methodology have been clarified, results will be discussed and implications for marketing managers will be also presented.

5.2. Theoretical models to analyze cultural differences

Individuals' attitudes towards GM food differ across countries (Zechendorf 1998): the US market, for example, has traditionally been indicated as one of the most supportive of transgenic breeding. High levels of skepticism, by contrast, have been detected in Europe, where GM food imports, as well as GM seed imports, have encountered tough opposition (primarily in Italy, Greece, and Austria). Such divergent opinions indicate that cultural factors – such as norms, values, or religion – could play a critical role in the explanation of people's disposition towards GM food. National communities may indeed be more or less inclined to deal with risky events (Hofstede 1983), or may tend to value economic growth and technological progress over environmental protection (Siegrist 1999).

To appraise the "cultural distance" between different social contexts, Hofstede (1983, 1984) proposed a model based on five main sociological and psychological dimensions denoted as: *Masculinity/Femininity*, referring to the extent to which male and female roles in a certain society are maintained distinct, or, conversely, tend to overlap; *Uncertainty avoidance*, reflecting a group's rejection of indefinite situations and its need for explicit rules; *Individualism/Collectivism*, expressing the extent to which a group's members care about social cohesion and are inclined to interact with each other; *Power distance* representing the level of inequality perceived by the members of a group with respect to the internal distribution of wealth and socio-political Power; *Long term/Short term orientation*, reflecting a nation's tendency to embrace a forward-looking perspective or, alternatively, to focus on the present and to prize values like preservation of traditions and established relationships (Hofstede 1983). According to Hofstede's (1983) theory, national communities may focus on personal success and materialistic goals or, conversely, may be more concerned about the welfare of their members and stress the need for cooperation. Northern European societies, for example, have been found to rely mostly on values like helpfulness and reciprocal support; whereas Southern European groups, in many cases, have been shown to deal mainly with competitiveness and attainment of personal success (Zechendorf 1998).

Human *values* are among the main factors which contribute to delineate the cultural profile of a group (Hofstede 1983) and can be defined as abstract beliefs which "transcend specific situation" and act as "guiding principles" in one's life (Schwartz and Bilsky 1990: 878). Examining data gathered in 67 countries, Schwartz (1992) has proved that values are shared worldwide and national communities may be distinguished according to the degree of importance they attribute to specific typologies of values. Precisely, Schwartz's (1992) theory postulates the existence of 56 different values which can be grouped according to their similarity in ten meta-dimensions referred to as "value domains". The ten domains have been shown to interact with each other through a series of relationships of compatibility and incompatibility. This system of interconnections has been represented on a two dimensional space as an elliptical structure (see Figure 5.1) within which compatible values have been arranged in adjacent positions (e.g., *Stimulation* and *Hedonism*), whereas incompatible values have been located in opposite settings (e.g., *Universalism* and *Power*). By grouping domains according to their compatibility or incompatibility, four meta-dimensions have been, finally, distinguished: (*i*) *Self-transcendence*, referring to the endorsement of altruistic goals, such as the welfare of nature and human beings, and within which the domains *Benevolence* and *Universalism* are located; (*ii*) *Self-enhancement*, reflecting the pursuit of selfish interests like the control over material resources and other people, and encompassing the domains *Achievement* and *Power*; (*iii*) *Openness to change*, referring to the search for variety and change, and grouping the domains *Self-direction*, *Stimulation*, and *Hedonism*; (*iv*) *Conservativism*, reflecting individuals and groups' need for certainty and stability, and referring

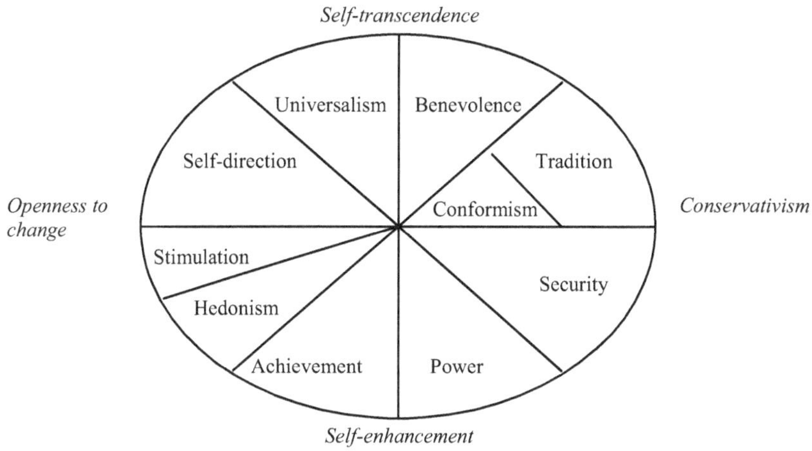

Figure 5.1. The universal structure of human values (Schwartz 1992).

to the domains *Conformism*, *Security*, and *Tradition*. Depending on their educational background, as well as their social life experiences and contingent situations (e.g., the geographic location), a group's members will be inclined to prioritize individualistic or collectivistic values, will be supportive of political, social, and technological changes, or, conversely, will regard security and traditions as basic principles in life.

These four different orientations have been addressed by Tsioumanis *et al.* (2004) to examine dissimilarities among European consumers' acceptance of GM food. They argue that a negative disposition towards GM food products is connected with the pursuit of the values falling within Schwartz's (1992) *Conservativism* and *Self-transcendence* meta-dimensions, since respect for tradition, need for security, and perception of sense of responsibility towards nature and other people may induce individuals to disapprove of the application of biotechnology to food production. Conversely, the attitudes of groups endorsing the values relative to the *Self-enhancement* meta-dimension (Figure 5.1, *above*) could be expected to be less negative, as "prioritizing personal goals over in-group goals may lead to actions that neglect the needs of future generations and/or the environment" (2004: 8). Finally, the authors sustain that groups attaching prominent importance to the values relating to *Openness to change* are more likely than others to rely on experts' understanding of gene technology when forming their opinions.

5.3. The nature of attitude towards GM food

Values do not only characterize national groups' beliefs, but also represent the drivers of a person's behavioral choices (Tsioumanis *et al.* 2004): they influence individuals' evaluations of reality and impact, through the formation of favorable/unfavorable attitudes towards given objects (e.g., beliefs, habits, products, etc.), on the decision to perform a certain behavior (Homer and Kahle 1988). An *Attitude* represents a tendency "to evaluate a certain entity with some degree of favorability or unfavorability" (Eagly and Chaiken 2007: 583) and results from an individuals' beliefs with respect to a particular object (Ajzen 1991).

As regards GM food, however, it has been observed that essential characteristics of such a typology of products – like the nutritional content, or the long-term effects deriving from their assumption – are unknown to the majority of consumers (Bredhal 2001). As a consequence, it could be argued that people's disposition towards GM food is more likely to reflect an overall judgment of biotechnology, or technology in general, rather than of the product by itself. From a theoretical viewpoint, this means that individuals will develop their attitudes towards the examined product according to a top-down process (Grunert *et al.* 2004) and, thus, their opinions will be more or less favorable according to their propensity to support economic growth and scientific progress (i.e., *Anthropocentrism*) or, from the opposite point of view, to preserve the integrity of nature and ecosystems (i.e., *Ecocentrism*) (Siegrist 1999). The top-down approach to the analysis of individuals' attitudes also suggests that dispositions towards abstract concepts, like technology or nature, are rooted in people's personal systems of values, which indirectly orientate their perceptions of innovative products, like GM food, and their beliefs about risks and benefits arising from technical innovations (Bredhal 2001; Grunert *et al.* 2004). A negative association has indeed been shown in reference to the link between *attitude* towards biotechnology and the tendency to favor the protection of nature *per se*, rather than its usefulness to humans (Siegrist 1998). By contrast, favorable *attitude* towards technology and trust in scientific progress have been proved to exert a positive influence on biotechnology acceptance (Siegrist 2003).

Research on GM food acceptance has focused on values to gain insight into consumers' dispositions towards these products. While attitudes are expected to change in valence (e.g., positive-negative, favorable-unfavorable) from person to person, values affect behavioral decisions to a greater or lesser extent, according to the level of importance people attach to them (Honkanen, Verplanken and Olsen 2006) and, as they are shared at the national level, contribute to define the identity of a group (Schwartz 1992). Greek consumers' attitudes toward GM food, for example, have been demonstrated to be negatively affected by the values pertaining to Schwartz's (1992) *Universalism* domain (e.g., *Equality*, *Environmental protection*, *Harmony with nature*) (Springer *et al.* 2005). Conversely, a sample of Dutch student's attitudes proved to be positively influenced by the values collapsing into the *Power* value domain (e.g., *Authority* and *Social power*) (Dreezens *et al.* 2005).

The careful inspection of the relationships between values and attitudes is, therefore, considered of crucial importance to determine people's willingness to buy GM food: stemming from a complex set of beliefs (which can be more or less directly related to these particular products), attitudes towards GM food are supposed to connect personal values to purchasing intentions and provide useful information to predict not only consumers' expectations about such products, but also behavioral choices (Honkanen and Verplanken 2004).

5.4. Aim and objectives

This study attempts to analyze *attitude* formation with regard to GM food products from an intercultural perspective. A research by Honkanen and Verplanken (2004) was replicated in Italy to achieve such a purpose. Administering their survey to a sample of Norwegian students, the aforesaid authors showed a significant relationship between *attitude* towards GM food and purchase *intention*. A similar result was achieved by Bredhal (2001), who analyzed data collected in four different European countries. Thus, in the light of these findings, it seems plausible to hypothesize that:

H1: GM food purchase intention is affected by individuals' attitude toward such products.

As stated above, the *attitude* a person holds towards a given object is supposed to mirror a certain system of values and to affect, in turn, this person's behavior (Homer and Kahle 1988). People prizing values like *Environmental protection* or *Helpfulness*, for example, can be expected to hold a favorable *attitude* towards recycling or donating to charity and, as a consequence, to behave accordingly in the majority of cases (Verplanken and Holland 2002). The validity of such a consideration has been confirmed by a series of studies on fair trade consumption (Doran 2009), willingness to behave in an environmentally friendly manner (Karp 1996), and to consume organic food products (Grunert and Juhl 1995).

Taken as a whole, these studies reveal that the pursuit of the values located in Schwartz's (1992) *Self-transcendence* meta-dimension (Figure 5.1, *above*) motivate individuals to engage in ethical behaviors. Such values, in fact, "are likely to be congruent with an ethical disposition as these are concerned with the harm that ethical transgression can do to others" (Steenhaut and Van Kenhove 2006: 141). In reference to GM food, the values pertaining to the *Universalism* domain, in particular (see Figure 5.1, *above*), have been showed to be inversely related to *attitude* and buying *intention* (e.g., Honkanen and Verplanken 2004). This implies that the more individuals are concerned about the risks that the diffusion of GMOs may involve for humankind and the environment, the less supportive they will be of their development.

The decisional process underlying the intention to purchase GM food has been modeled by Honkanen and Verplanken (2004) as a "causal flow" departing from personal values and feeding into buying *intention*. Such a decisional pattern, in the case of sustainable consumption choices, proved to arise in the expected direction, at least in the short term since "values changes at the societal level are caused by new generations replacing old ones" (Thøgersen and Ölander 2002: 607). For these reasons, Honkanen and Verplanken (2004) reputed a *mediation model* (Baron and Kenny 1986) to be suitable for the purpose of their study. Mediation analysis enables sociological and psychological researchers to gain an insight into many behavioral processes by distinguishing the *direct effect*, exerted by an alleged predictor variable on a dependent variable, from the *indirect effect*, due to the mediating action of a third variable, referred to as "mediator" (MacKinnon, Fairchild and Fritz 2007). The sum of both the direct and the indirect effects is posited to return the measure of the whole predictive effect shaped by the independent variable on the dependent variable (*total effect*).

Honkanen and Verplanken (2004) showed that the endorsement of the values encompassed in Schwartz's (1992) *Universalism* value domain negatively influence the *attitude* towards GM food which, in turn, was found to mediate the effect of *Universalism* on the *intention* to buy these products. As a consequence, it is here supposed that:

H2: Attitude towards GM food is negatively affected by the values pertaining to the Universalism domain.

H3: GM food purchase intention is indirectly and negatively affected by the values pertaining to the Universalism domain, with attitude acting as mediator.

The conclusions reached by Honkanen and Verplanken (2004) appear to be also supported by considerations of cultural order. Their survey was carried out in Norway, a country which, according to Hofstede's (1983) model, is characterized by a high degree of *Femininity*. That is, social roles of males and females tend to overlap, and values like solidarity, cooperation, and caring for the environment are highly regarded. Italy, by contrast, is ranked as one of the most *Masculine* countries among those examined in Hofstede's (1983) study; consequently the distinction between gender roles is more evident and material goals, such as personal accomplishment and achievement, are prioritized (Hofstede 2001). It could be expected, therefore, that *attitude* towards GM food and the subsequent purchase *intention* may be positively affected by values like *Social power* or *Authority* which, as previously explained, are grouped in Schwartz's (1992) *Power* value domain. More precisely, since attitudes towards GM food have been supposed to directly affect GM food purchase *intention*, the *Power* value domain is expected to exert an indirect effect on purchase *intention* that will be transmitted by individuals' attitudes toward GM food. Hence, the validity of the following statements is predicted:

H4: Attitude towards GM food is positively affected by the values pertaining to the Power domain.

H5: GM food purchase intention is indirectly and positively affected by the values pertaining to the Power domain, with attitude acting as mediator.

In accordance with the hypotheses formulated by Honkanen and Verplanken (2004), and the top-down *attitude* formation process postulated by Grunert *et al.* (2003), in this study, attitudes towards GM food are, therefore, supposed to be responsible for transferring the effect exerted by personal values (taken as independent variables) on purchase *intention* (taken as the dependent variable) and to act as a mediating variable.

5.5. Methodology

The rationale for this study is to throw light on cultural factors affecting consumers' perception of GM food products. To obtain data comparable with those of Honkanen and Verplanken (2004), items used by the aforesaid authors in their research were inserted in a questionnaire that was administered to a sample of university students, as in the case of the original survey. Details concerning the sampling procedure and the measures adopted are presented in the following sections.

5.5.1. Sampling procedure

A sample of 251 Italian subjects was drawn from the student population of a medium-sized university in Lecce (Italy). The majority of the respondents were female (70%) and their average age was 22 (standard deviation = 2.87).

5.5.2. Measures and questionnaire

Data were gathered through a paper-and-pencil questionnaire proposing a brief definition of "genetically modified food" in the foreword and consisting of a series of close-ended questions which were developed from the survey instrument previously administered in Norway by Honkanen and Verplanken (2004). Items relating to *attitude* towards GM food, GM food purchase *intention*, and personal values were translated into Italian and structured into three distinct sections.

Attitudes were assessed asking participants to answer the following question: "I think that eating GM food products is…". Answers were collected by means of three items measured on a bipolar semantic differential scale, ranging from 1 to 7, and marked at the extremes with the following adjectives: "foolish-wise"; "unreasonable-reasonable"; and "negative-positive".

To ascertain the level of purchase *intention* relating to GM food products, respondents were requested to express their level of agreement to the following three statements: "Are you going to buy GM food?"; "Are you going to buy GM food if the quality is better compared to traditional food?"; "Are you going to buy GM food if it is cheaper than traditional food?". Answers were measured on a 7-point categorical scale labeled with the word "Maybe" at the mid-point, and with the expressions "Certainly not" (corresponding to 1) and "Certainly" (corresponding to 7) at the end-points.

Finally, personal values were measured asking subjects to fill in the *Schwartz Value Survey* (SVS, Schwartz 1992), a 9-point scale requiring respondents to rate the importance of fifty-six different values (e.g., *Equality*, *Enjoying life*, etc.). The extremes of the scale were marked with the expressions "Opposed to my values" (corresponding to -1) and "Of supreme importance" (corresponding to 7), while the scale mid-point was labeled with the word "Important".

5.6. Results

Having checked the internal reliability of the scales employed to collect data, the basic descriptive statistics were computed. Then, Baron and Kenny's (1986) procedure was followed to test the research hypotheses. Details about the survey results are presented in the following sections.

5.6.1. Reliability analysis

Cronbach's α coefficients were calculated for each of the studied constructs to appraise the level of inter-item correlation (Table 5.1). *Attitude* towards GM food proved to be the most reliable scale (Cronbach's α = .90), whereas a satisfactory level of internal consistency was found for *Universalism*, *Power*, and *intention* measures (Cronbach's $\alpha \geq .70$). These results were in part confirmed by a series of *principal component analyses* performed with respect to each construct (Table 5.1). Items used for measuring each variable proved to be reducible to a single dimension accounting for more than 60% of their variance. As expected, the highest percentage of explained variance was observed for the *attitude* scale (84%).

Variables	*Cronbach's α*	*Explained variance*	*Eigenvalues*
INT	.785	71.021	2.131
ATT	.905	84.362	2.531
UNI	.793	62.616	3.452
POW	.758	67.590	2.028

Note: *N* = 251; INT = Intention; ATT = Attitude; UNI = Universalism; POW = Power.

Table 5.1. Results of internal reliability analysis.

5.6.2. Descriptive analysis

Data were synthesized by calculating means and standard deviations of the examined variables. Results reveal that participants' attitudes towards GM food were mostly negative (Table 5.2): a mean value of 2.571 was ascertained on a scale ranging from 1 to 7. Out of 251 respondents, 207 (82%) scored their average *attitude* below the scale midpoint. Equivalent results were detected for the purchase *intention* (M = 2.527, SD = 1.321). The majority of participants considered the values relative to *Universalism* as important principles in life (M = 5.356, SD = .949), whereas the average level of importance attached to the values inherent in the *Power* domain was rather low: a mean value of 1.771 was reached on a 9-point categorical scale (SD = 1.580) and out of the 251 respondents, 56 (22%) rated *Power* by using the scores ranging from "Important" (corresponding to the scale midpoint) to "Of supreme importance" (corresponding to the positive extreme of the scale).

5.6.3. Hypotheses testing

To assess the existence of linear relationships between the examined variables, Pearson's correlation coefficients were computed. The resulting matrix, represented in Table 5.3 (*below*), denotes the presence of moderate levels of correlation within the constructs. As expected, the association between *attitude* towards GM food and purchase intention was stronger than those ascertained in the other cases (r = .657).

Variables	*M*	*SE*	*SD*
INT	2.527	.083	1.321
ATT	2.571	.075	1.201
UNI	5.356	.060	.949
POW	1.771	.099	1.580

Note: *N* = 251; INT = Intention; ATT = Attitude; UNI = Universalism; POW = Power.

Table 5.2. Descriptive statistics.

	INT	ATT	UNI	POW
INT	1.000	-	-	-
ATT	.657*	1.000	-	-
UNI	-.128*	-.109	1.000	-
POW	.491*	.376*	-.024	1.000

Note: $N = 251$; INT = Intention; ATT = Attitude; UNI = Universalism; POW = Power; * = Significant at the .05 level.

Table 5.3. Pearson's correlation coefficients.

Since all correlation coefficients were shown to be lower than .70, the likelihood of multicollinearity conditions proved to be minimal. It should be noted that the buying *intention* transpired to be negatively correlated with *Universalism*, whereas the relationships between *Universalism* and the remaining variables – i.e., *attitude* and *Power* – were found to be not significant ($p > .05$).

The first research hypothesis (H1) was tested entering *attitude* and *intention* in a least square regression equation as the independent and the dependent variable, respectively (Table 5.4). The fit of this model was adequate ($R^2 = .432$) and, therefore, it was possible to confirm that *attitude* towards GM food represents a proximal determinant of subjects' *intention* to purchase this type of products ($\beta = .657$; $p < .005$).

Baron and Kenny's (1986) multi-step approach to mediation analysis was subsequently implemented to simultaneously test the hypotheses concerning the incidence of *Universalism* (H2 and H3), as well as that of *Power* (H4 and H5), on respondents' decisions. However, as regards Hypotheses H2 and H3, the weak correlations

Variables	*B*	*Beta*	*t*	*p*
Constant	.669		4.488	.001
ATT	*.723*	*.657*	*13.749*	*.001*

Note: Dependent variable: Intention; $N = 251$; $R = .657$; $R^2 = .432$; Adjusted $R^2 = .429$; ATT = Attitude.

Table 5.4. Regression of purchase *intention* on *attitude* towards GM food products.

previously detected among *Universalism*, *attitude*, and *intention* (see Table 5.2, *above*) suggested that one firstly check the applicability of the model by regressing the dependent variable (i.e., *intention*) on the predictor (i.e., *Universalism*). Such an analysis, which tested a precondition for a mediation process to occur, yielded a poor fit and led, therefore, to reject the hypotheses in question.

To test Hypotheses H4 and H5, *Power* and *intention* were posited to act as independent and dependent variable, respectively, while the relationship between these two constructs was assumed to be mediated by *attitude*. A series of independent linear regression equations were run (Table 5.5): firstly, *Power* was demonstrated to be a significant antecedent of *intention* (*Step 1*: $R^2 = .141$; $\beta = .491$; $p < .005$), thus confirming Bredhal's (2001) considerations about the influence exerted by personal values in the decision-making process leading to purchase GM food. Secondly, *attitude* was demonstrated to be significantly affected by the values relating to the *Power* domain (*Step 2*: $R^2 = .247$; $\beta = .376$; $p < .005$); as a consequence, it was possible to confirm the validity of Hypothesis H4. Lastly, *intention* was regressed on *Power* and *attitude*, considered as independent and control variable, respectively (*Step 3*). As in this case the link between *Power* and *intention* ($\beta = .284$; $p < .005$) proved to be weaker than the one ascertained by the first regression equation ($\beta = .491$; $p < .005$), it can be

Steps	*Variables*		*B*	*Beta*	*t*	*p*
1.		Constant	1.800		16.448	.001
	Independent:	*POW*	*.410*	*.491*	*8.890*	*.001*
	Dependent:	INT				
2.		Constant	2.065		19.514	.001
	Independent:	*POW*	*.285*	*.376*	*6.397*	*.001*
	Dependent:	ATT				
3.		Constant	.550		3.891	.001
	Independent:	*POW*	*.238*	*.284*	*5.871*	*.001*
		ATT	*.605*	.550	*11.366*	*.001*
	Dependent:	INT				

Note: $N = 251$; $R^2_{(1)} = .141$; Adjusted $R^2_{(1)} = .138$; $R^2_{(2)} = .247$; Adjusted $R^2_{(2)} = .244$; $R^2_{(3)} = .436$; Adjusted $R^2_{(3)} = .432$; INT = Intention; ATT = Attitude; POW = Power.

Table 5.5. Application of mediation analysis: result summary.

argued that a *partial* mediating effect occurred. That is, only a part of the influence shaped by *Power* on *intention* was transferred by *attitude* (see Table 5.5, *above*). The whole predictive effect of *Power* on *intention* (*total effect*) proved, therefore, to be transmitted in part directly (*direct effect*), and in part indirectly (*indirect effect*), via consumers' *attitude* towards GM food (β = .550; $p < .005$). Referring to the beta weights presented in Table 5.5, the total effect was decomposed as indicated in the following equation: *total effect* = [*direct effect* + (*indirect effect*)] = [.284 + (.376 × .550)] = .491. Consistently with Baron and Kenny's (1986) specifications, Hypothesis H5 was, then, accepted.

5.7. Discussion

While Honkanen and Verplaken (2004) demonstrated that a sample of Norwegian students' disfavor towards the use of GMOs is embedded in a sense of responsibility towards nature, other people, and future generations (*Universalism*), such a mental disposition, for Italian respondents, proved to possess weak motivational force. Participants' negative attitudes towards GM food were shown to stem from a tendency to attribute minor importance to mastery of nature and social relations (*Power*). This study demonstrated, indeed, that the endorsement of the values grouped in the domain referred to as *Power* in Schwartz's (1992) *Universal theory of values* affects not only individuals' disposition towards GM food, but also their potential purchase *intention*.

To date, the relationship between *Power* and consumers' behavior has been examined by several researchers: Doran (2009), for example, documents the presence of an inverse correlation between the values located in this domain and purchase of fair trade products, whereas Karp (1996) detects a negative influence of the whole *Self-enhancement* meta-dimension on people's willingness to engage in environmentally friendly behaviors like recycling and purchase of eco-labeled products. Considered together, these studies suggest that the propensity to attach major importance to the values located in the *Self-enhancement* domain of Schwartz's (1992) representation of human values (see Figure 5.1, *above*) is likely to determine the adoption of unethical patterns of behavior. As Steenhaut and Van Kenhove (2006: 141) point out, in fact, "*self-enhancement* goals are likely to be congruent with an unethical disposition, because unethical consumer practices involve promoting personal interests, with little or no regard for others". *Self-transcendence* values, by contrast, have been shown to motivate consumers to care about the ethical reflections of their behavioral choices (cf. Steenhaut and Van Kenhove 2006). As outlined above, self-transcendent persons are likely to recycle waste, purchase eco-friendly products (Karp 1996), and, as Honkanen and Verplanken (2004) demonstrated, disapprove of the employment of GMOs in food production.

Results similar to those ascertained in this study were obtained by Dreezens *et al.* (2005), who also showed that individuals considering values like *Authority*, *Social power*, or *Social recognition* as important principles in their life are likely to hold a favorable disposition towards biotechnology and GM food and, as a consequence, may be willing to include GM food products in their regular diet. Such a conclusion appears congruent with empirical findings by Finucane *et al.* (2000), who investigated how, in the USA, different ethnical groups tend to perceive cultivation of GM crops and other forms of food-related risks (e.g. "mad-cow" disease, the presence of pesticide residues in fruits and vegetables, etc.). In their study, white males, compared to African-American, Hispanic, and Asian groups, were found to be less pessimistic about the development of GM food crops. Specifically, white males proved to perceive a lower level of risk in biotechnology applications and to support, in the majority of cases, a hierarchical and individualistic organization of society.

Aspects like control over events and individuals, as well as low risk perception, have attracted the interest of anthropological and psychological research. Specifically, dominance over other people and material resources has been considered as an expression of a *Masculine* view of the world (Fiddes 1989), that is an orientation towards materialistic goals and personal achievement. Whereas, risk perception has been supposed to be inversely related to the possibility to control uncertain events (e.g., Trucker, Whaley and Sharp 2006): the higher the control subjects believe they can exert over uncertain situations, the lower may be their perception of risk. From this twofold perspective, the divergence between Norway and Italy, seems noteworthy. By measuring *Masculinity* (MAS) on a scale of zero to 100, Hofstede (2009) found indeed that such a dimension for Norwegian culture (as well as the culture of the whole of Scandinavia) reaches a lower level ($MAS_{NO} = 8$) than that detected for Italian society ($MAS_{IT} = 70$). Moreover, a significant positive correlation between the *Masculinity* dimension and Schwartz's (1992) individualistic meta-dimension (i.e., *Self-enhancement*) was documented by Ng, Lee and Soutar (2007). The two national communities also differ in their tolerance for risk: Hofstede's (2009) *Uncertainty Avoidance Index* (UAI), indeed, suggests that unstructured and unpredictable situations determine a higher sense of discomfort in Italy ($UAI_{IT} = 75$), rather than in Norway ($UAI_{NO} = 50$). This finding represents a possible explanation of the slight difference detected between the two countries in terms of rejection of biotech products: Norwegian respondents' attitude towards GM food (mean value = 3.15) were not as negative as that of Italian respondents (mean value = 2.57); whereas the proportion of subjects unwilling to consume GM food was moderately smaller in Norway (70.5%) than in Italy (82%).

5.8. Implications

Currently, safety and healthiness constitute the critical factors that deter the majority of consumers from accepting innovative products and new processing technologies (Gaskell *et al.* 2004). This study, however, in line with a growing stream of research, also showed that opposition to the development of the market for GM food products is embedded in consumers' basic beliefs, that is in their personal systems of values. Traditionally, communication campaigns against the liberalization of GM crops have been centered on the principles of environmental protection and preservation of the ecological balances. According to the results of this study, emphasis should also be put on the negative sides of humans' attempt to "master" natural resources and "control" biological processes. Advertisers might call attention to the likelihood of unexpected reactions and irreversible alterations as the most detrimental aspects connected with the introduction of highly sophisticated technological innovations. They could also address public commitment with the ethical matters arising from biotechnological development. Indeed, as shown by Rozin (2005), compared to domestication, physical transformations, and other manipulations progressively introduced by humans, genetic engineering is perceived as highly destructive of natural orders.

On the other hand, the analyses performed in the present study also reveal that a minority of individuals, who tend to recognize *Authority* and *Social recognition* as distinctive characteristics of their personality, may hold a favorable *attitude* towards novel production methods and may be more likely than other people to appreciate specific food features developed through genetic experimentation. Consequently, it could be predicted that the potentialities of biotechnologies could be exploited by food producers to offer products endowed with properties like longer shelf-life, enhanced nutritional content, or shorter cooking-time in order to suit the needs of emergent niche markets (cf. Canavari and Nayga 2009). However, as Frewer (2004) observed, until the improvement of food healthiness and safety, and the achievement of broader objectives like a reduction of the impact of human activities on nature remain doubtful, public resistance to the introduction of GMOs in the agro-food industry will continue to be strong.

Advertisers and policy-makers should not neglect the role of cultural diversity, either. The present study has shown, in fact, that although consumers living in different countries may hold analogous negative attitudes towards GM food, such dispositions may stem from different convictions, according to the individualistic/altruistic orientation of the different nations. Thus, the distinctive characters of each national community should be taken into consideration to develop tailored communication policies appealing to consumers' sensitivity to the consequences deriving from the introduction of GMOs. Such an approach could stimulate public interest for the ethical and social issues arising from the topic of genetic manipulation, and may provide consumers with clear arguments to adopt a precise position in the debate about the applications of biotechnology.

5.9. Limitations and future research

The first limitation of this study regards the measurement of participants' attitudes towards GM food. It has been observed that bipolar scales like the one used in this study cannot provide an effective detection of individuals' judgments on controversial topics like genetics and biotechnology (Portinga and Pidgeon 2006). Although consumers have traditionally been classified as "favorable" or "unfavorable", many of them are unable to assume a clear position in the debate that surrounds the topic of GM food. In this survey, for example, 24% of the respondents expressed a neutral *attitude* towards these products. Measures of attitudinal ambivalence are needed, therefore, to gain a deeper insight into people's disposition towards the issue of biotechnology and its application to food production.

Additionally, it should be recognized that personal values are weekly supported by rational considerations, as they are commonly considered psychological dimensions which are mostly related to the emotional sphere of human personality (Maio *et al.* 2001). For this reason, actual conduct may often be inconsistent with a person's system of values since situational factors may prevent subjects from making coherent decisions. Also the extent to which an individual is motivated to behave in a way congruent with his/her basic principles (*value centrality*) needs, therefore, to be accurately investigated to corroborate the conclusions reached in the present study.

Finally, it should be pointed out that Ajzen's (1991) suggestions could be followed to draw more reliable information about individuals' behavioral intentions. Future research could, in fact, extend the model tested in this chapter by including measures of the pressure that others' judgments may exert on an individuals' decisions (*subjective norm*) and by considering the factors that are likely to deter consumers from purchasing GM food or, conversely, may positively impact on their buying intentions (*perceived behavioral control*).

5.10. Conclusions

The present chapter focused on individuals' disposition towards GM food and the relative willingness to consume these products. By replicating a study by Honkanen and Verplanken (2004) it was possible to compare the views of individuals living in two different nations – i.e., Italy and Norway – and to assess how cultural components affect people's behavioral choices. To understand the rationales underlying such choices, the interactions among some basic antecedents of human behaviors, namely *attitudes*, *intentions*, and *personal values* were addressed.

As in the original study, participants' attitudes towards GM food were found to be predominantly negative. However, while Honkanen and Verplanken (2004) showed that Norwegian respondents' negative dispositions are connected with a strong sensitivity towards nature and others' well-being, such a relationship, for Italian respondents, was found to be weak. GM food rejection, in this study, proved to be associated with a tendency to consider values like *Authority* and *Social power* as negligible principles in one's life. On the other hand, the pursuit of such values proved to positively affect the attitudes held by a small portion of the sample and, as a result, made it possible to identify potential consumers of GM food.

From a theoretical perspective, the presence of a linkage between personal values and both respondents' attitudes and behavioral intentions suggests that GM food consumption, and food consumption in general, can be conceived, as a whole, as expressions of individuals and groups' identities (cf. Colacchio, Mastrangelo and Zarantonello

2008). From an operational perspective, it could be argued that the analysis of individuals' value systems can provide decision-makers with useful information about the potential demand for new products, like GM food, and help them detect psychological factors that may hamper the development of new market segments. Moreover, by analyzing values at the national level, marketers may also gain awareness of those determinants rooted in the cultural profile of the different nations which contribute to define a groups' tolerance for uncertainty and high risk situations, as well as the ethical principles embraced by its members.

Alessandro M. Peluso

6. Determinants of consumers' acceptance of fuels from biological materials and hydrogen transportation technologies[1]

6.1. Introduction

For several years now, national governments have placed energy and environmental issues high among the priorities of their political agenda. They have been investing increasing efforts in seeking to secure a sustainable energy supply and to contribute to the mitigation of the atmospheric pollution that causes global warming (Edwards *et al.* 2008). Whereas the need for energy security stems from the necessity for many developed countries to reduce their dependence on foreign oil suppliers, their governments are often required by international agreements – such as the Kyoto Protocol in 1997 – to develop environmental protection programs, aimed at reducing greenhouse gas emissions (Dunn 2002). In particular, the transport sector has witnessed a continuing increase in the number of vehicles in circulation, probably due to the growing population worldwide and to an enhanced mobility of people. This phenomenon has been identified as one of the main causes of greenhouse gas emissions (Puppán 2002). The consequent augmentation of atmospheric pollution makes it reasonable to suppose that the use of petroleum-based fuels in civil transportation is unsustainable in the long run. This scenario may be further worsened by other macro-economic factors related to the high volatility of crude oil prices, as well as to the limited oil reserves available, which are under a quasi-monopolistic control of a restricted number of states

1 The Author would like to thank Prof. Gianluigi Guido for his comments on the draft of the manuscript.

(such as those belonging to the Organization of the Petroleum Exporting Countries, OPEC). A common strategy adopted by policy makers to address these issues consists in encouraging the development of renewable energy sources and transportation fuels such as *bio-fuels* (e.g., Baka and Roland-Holst 2009) and *hydrogen* (e.g., Zhao and Melaina 2006), which are commonly considered as among the most promising alternative energy sources to traditional fossil fuels.

Bio-fuels, in particular, refer to any solid, liquid or gaseous fuel derived from biomass materials including food products, such as corn, soybean and other crops (Pahl 2005). Lands and other resources are often used for energy production, thus giving rise to a paradigmatic shift from a pure agricultural model, based on the use of land for cultivating food crops, to an "ergocultural" one, based on the use of land for energy purposes (Mathews and Goldsztein 2009: p. 327). This raises ethical concerns that are the subject of an interesting debate among researchers (cf. Thompson 2008), insofar as bio-fuel production can affect negatively food security and availability, that is, to make food products accessible to the population of a social system – such as a geographical region or a nation (cf. Srinivasan 2009). This phenomenon may occur not only through the use of traditional food crops as raw materials in energy production, but also through the diversion of tangible and intangible resources (e.g., lands, financial resources, and skills) away from agro-food production. This may lead to an increase in price levels of commodities in international markets, thereby exerting a negative impact on food supplies – as also acknowledged internationally by the Food Agriculture Organization (2008).

Hydrogen has been suggested as another viable alternative to conventional fuels that could be used in civil transportation (e.g., Johnston, Mayo and Khare 2005). It is not an energy source, but a "carrier", that is, a secondary form of energy that must be produced from other primary sources (Balat 2008). Whereas bio-fuels are deemed to be a short-term option to petroleum-based fuels, hydrogen is still considered as an emerging technology that could be available to the mass market only in the mid to long term (Ramesohl and Merten 2006). Hydrogen transportation technologies lag behind those of bio-fuel at both an economic and a technical level, due to specific barriers regarding production, storage, distribution, costing, as well as public acceptance of hydrogen technologies, based on the individual's perception of safety and risks connected with their use as an alternative fuel (McDowall and Eames 2006).

Individual perception of renewable energy sources and new transportation technologies are relatively neglected areas of research. No study on public perception and acceptance is available in the bio-fuel literature, whereas the sporadic studies carried out on hydrogen technologies are either conceptual (e.g., Schulte, Hart and Van der Vorst 2004) or merely explorative, without adopting any robust theoretical framework (e.g., Hickson, Phillips and Morales 2007). The present chapter aims, therefore, at investigating the cognitive beliefs consumers hold about bio-fuel and hydrogen transportation technologies. It examines potential users' perceptions and ethical evaluations of advantages and disadvantages, social pressures, and external factors potentially capable of determining the acceptance and adoption of these technologies. Although hydrogen seems to be less controversial than bio-fuels from an ethical perspective, as it is abundant in the natural environment and its production should not threaten food security, potential users could however be influenced by ethical considerations in their perception and evaluation of its derived transportation technologies like hydrogen cars. The next section will describe the characteristics of bio-fuels and hydrogen as potential alternatives to conventional fuels. The third section will illustrate a study on subjective beliefs associated with bio-fuel acceptance and adoption. The fourth section will report a second study on subjective beliefs related to hydrogen as a new transportation technology. The fifth section will provide a general discussion of results, and conclude the chapter by delineating theoretical and marketing implications.

6.2. Bio-fuels and hydrogen transportation technologies

Bio-fuels and *hydrogen* are among the most promising technologies that could contribute to reducing atmospheric pollution and the chronic oil crisis related to a strong dependence of Western developed economies on imported oil supplies. The next two sub-sections will illustrate the main characteristics of these technologies and their potential in the transportation industry.

6.2.1. Bio-fuels in the transportation industry

Bio-fuels are a controversial alternative to conventional transportation fuels, as the research in this field has shown contrasting results regarding the potential advantages and disadvantages deriving from their diffusion as transportation fuel. Demirbas (2007ab) considers bio-fuels as an environmentally sustainable alternative energy source capable of reducing greenhouse gas emissions and favoring local development. Yet, he also describes bio-fuels as an alternative that is still not a serious competitor to petroleum-based fuels, such as diesel oil and petrol, from an economic perspective. Contrasting evidence also emerges from the existing research on their environmental impact. Whereas Puppán (2002) shows that such an impact is, in general, more favorable than that of conventional fuels, Pimentel and Patzek (2005) demonstrate that this impact becomes negative when bio-fuels are produced from energy crops like sunflower, due to specific physical characteristics limiting the amount of oil that can be extracted from this crop in comparison with other species (such as soybean).

As a potential driver of local development in rural areas, recent studies have examined the impact of bio-fuel diffusion on specific domestic economies. For example, Gokcol *et al.* (2009) describe bio-fuels as an opportunity for the development of the Turkish economy. As regards the diffusion of bio-fuels in the US market, Morrone *et al.* (2009) stress the existence of specific economic barriers (e.g., energy costs associated with bio-fuel production) and socio-political ones (e.g., environmental costs associated with the loss of high-quality forests). In the European context, Bomb *et al.* (2007) analyze comparatively the consequences deriving from bio-fuel introduction in Germany and UK and provide a range of policy recommendations, including, among others: the importance of duty exceptions or reduction, for stimulating bio-fuel diffusion; the need for national governments to play a supportive role in developing bio-fuel industry; and the necessity of sustainability certification for bio-fuels and their by-products. In a recent study carried out on the Italian context, Russi (2008) points out that the production of bio-fuels, in general, and bio-diesel, in particular, is still economically inconvenient and environmentally unsustainable. Against marginal advantages in terms of urban pollution reduction and rural development there are relevant disadvantages from both an economic and an environmental point of view. From an economic perspective, bio-diesel production is a high-energy consuming activity, with an unfavorable input/output energetic ratio. Moreover, the existing technologies for bio-fuel production do not meet efficiency, efficacy, and profitability standards. From an environmental perspective, the diffusion of large-scale production systems may exert a negative impact on the environment, biodiversity, and natural habitats, as the cultivation of energy crops requires the use of intensive agricultural methods and increasing amounts of fertilizers (see also Koh 2007; Yang, Zhou and Liu 2009). These arguments contrast with the common conception of bio-fuels as a clean alternative to fossil fuels. Most importantly, they contrast with recent policies adopted in Europe by central and local authorities to encourage the adoption of bio-fuels by individual consumers and business organizations (cf. Frondel and Peters 2007).

Bio-diesel, technically defined as a vegetable oil methyl ester (cf. Demirbas 2007b), is absolutely the most common and widespread type of bio-fuel worldwide, and in European countries in particular (see Figure 6.1, *below*). In Europe, it represents around the 80% of total bio-fuel production: the European Union is the world leader in bio-diesel production, with Germany, France, and Italy being the largest producers (Pahl 2005). Moreover, the European Commission has been encouraging the diffusion of this renewable energy source among member states.[2]

---

2 The Directive 2003/30/EC was introduced with the intent to promote a progressive substitution of traditional fossil fuels (such as diesel oil, petrol, and the like) with bio-fuels, and bio-diesel in particular. The European Commission's (2006) communication strategies focused on specific advantages deriving from a large-scale adoption of bio-diesel: (*i*) the possibility of using blends of bio-fuels for transportation, without the necessity of modifying the structural characteristics of engines; (*ii*) the possibility of using the existing storage

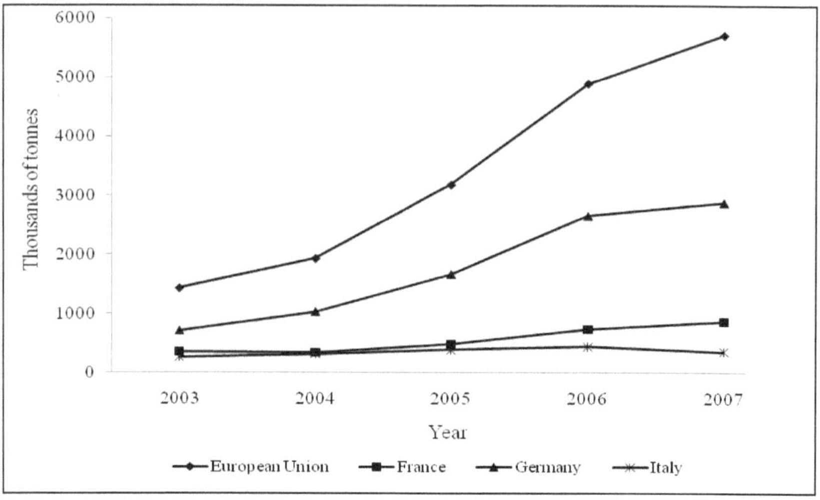

Source: European Bio-diesel Board 2009.

Figure 6.1. Bio-diesel production in leading European countries (2003-2007).

From this review, the existence of contrasting positions emerges, regarding the validity of bio-fuels, in general, and bio-diesel, in particular, as alternative environmentally-friendly energy sources to traditional fuels. For these reasons, national governments, car-makers (such as BMW, General Motors, and Honda), and energy companies (such as BP, Shell, and Chevron Texaco) are also committed to developing investment programs regarding other renewable sources such as hydrogen (Johnston, Mayo and Khare 2005; Solomon and Banerjee 2006).

---

and distribution systems (e.g., the same pumps used for diesel oil and petrol); (*iii*) the environmental neutrality of bio-diesel, that is, its ability to emit a lesser amount of greenhouse gases (e.g., carbon dioxide) during its combustion than that absorbed from energy crops while growing; (*iv*) their energetic efficiency, that is, the ability of these crops to produce a greater amount of energy than that required from growing them; (*v*) the possibility of diversifying the energy supply-chain in Europe; as well as (*vi*) the possibility of favoring the development of agriculture sector and a consequent increase in employment.

6.2.2. Hydrogen technologies in the transportation industry

Hydrogen is not a resource (like petroleum or bio-fuels), but an energy carrier (like electricity) that must be derived from a primary source through specific physical-chemical processes. An important advantage in this regard is that hydrogen can be produced from a variety of energy sources, including natural gas, methane and other hydrocarbons, as well as from renewable sources like nuclear, wind, and solar energy (Balat 2008). This means that, although hydrogen is often promoted as a "clean fuel" (Cherry 2004: p. 127), its production is not always free of greenhouse gas emissions. In this regard, Van Ruijven *et al.* (2008) stress that deriving hydrogen from fossil fuels rather than renewable energy sources is not a sustainable solution, as it causes a shift of pollutant emissions from the phase of the end-use of hydrogen transportation technologies to the first stage of the production process. In the transportation industry, the production of hydrogen from renewable sources represents a challenge for developed countries, including the European Union, which has planned to reach this goal in coming decades (Edwards *et al.* 2008). Another advantage stemming from the use of hydrogen in civil transportation pertains to its physical-chemical properties which enhance its efficiency in powering engines (e.g., the fact that it burns rapidly with a high effective octane number).

Against these benefits, there are serious disadvantages that represent barriers to the diffusion of hydrogen cars and other transportation technologies. According to McDowall and Eames (2006), the main obstacles to the development of a hydrogen economy regard: (*i*) the absence of hydrogen refueling infrastructures; (*ii*) the high costs associated with hydrogen production and delivery; and (*iii*) the immaturity of the related technologies used in the manufacture of hydrogen cars and their on-board storage systems (i.e., fuel cells). Other researchers agree with this assessment. Blanchette (2008), for example, indicates similar barriers to the diffusion of these technologies, but also stresses how the transition to the hydrogen economy needs active involvement on the part of national governments. In accordance with this conclusion, Farrell, Keith and Corbett (2003) posit that, for hydrogen to become of common use in civil transportation, actual market conditions should change and central authorities should adopt appropriate energy policies. On the other hand, Zerta *et al.* (2008) agree on the existence of such barriers, but provide a more optimistic view by maintaining that such a transition is feasible and suggest that hydrogen is among the best alternatives to fossil fuels in civil transportation.

These barriers to the diffusion of hydrogen transportation technologies open up challenges for researchers and policy makers. Edwards *et al.* (2008) point out that, for these technologies to become sustainable and available to the mass market, hydrogen production, distribution and storage systems should be improved, thereby indicating specific challenges regarding the needs to: (*i*) lower the cost of hydrogen production to a level comparable with other fuels; (*ii*) develop environmentally-friendly hydrogen production systems; (*iii*) increase safe and efficient hydrogen delivery, distribution, and onboard storage systems; and (*iv*) reduce costs and improve the performance of hydrogen-powered vehicles.

Whereas these challenges have attracted a growing research interest (see Gupta 2008, for a review), that regarding the need to address the possibility that consumers hold inaccurate perceptions of hydrogen technologies has remained relatively neglected. Few researchers have stressed the importance of such a perception. Cherry (2004) was the first researcher to suggest that consumers' concerns about safety and risks associated with hydrogen transportation technologies could impede the latter's acceptance and diffusion. Johnston, Mayo, and Khare (2005) further stress the importance of this aspect and place the need to improve the public perception of hydrogen technologies among the main challenges for policy makers and automotive companies. To date, few other studies (Dinse 1999, 2000; Ludwig-Bölkow-Systemtechnik 1998) have explicitly investigated consumers' perceptions and acceptance levels of hydrogen-powered cars. They have found that individuals do not display a negative attitude towards this alternative fuel and only sporadically associate it with tragic accidents (e.g., the Hindenburg disaster in 1937). In this research stream, Schulte, Hart, and Van der Vorst (2004) develop a conceptual model and conclude that marketing investments are important for a new technology like hydrogen to be accepted by consumers. Efforts should be spent on educational programs, aimed at informing potential users of the characteristics and benefits associated with this technology. Investments should also be made in developing communication campaigns, designed to project a positive image of hydrogen as a transportation fuel. Specific marketing initiatives should be adopted in order to provide potential consumers with a direct consumption experience with the product (i.e., hydrogen-fuelled cars). These conclusions are further supported by Hickson, Phillips and Morales (2007), who show that consumers' perceive hydrogen-powered vehicles as an environmentally-friendly technology and indicate consumers who are sensitive to environmental issues as the main target of companies' marketing strategies.

In light of the relevance of issues like energy security and global warming, it is possible to hypothesize that the diffusion of renewable energy sources may be crucial to the well-being of future generations (cf. Elliott 2000). Yet, the adoption of renewable transportation fuels like bio-fuels and hydrogen, in alternative to fossil fuels, is likely to be guided not only by their perceived consequences – in terms of advantages and disadvantages – but also by the ethical principles of potential users.

The ethical valence of bio-diesel and hydrogen as energy sources clearly emerges from two recent studies by Peluso (2008), and Guido *et al.* (2008), respectively. These studies were aimed at identifying subjective beliefs driving the potential users' evaluation and acceptance of these emerging technologies. In this chapter, their qualitative data are re-examined and results discussed through the lens of the ethical issues (such as environmental protection) that consumers may consider while evaluating these transportation technologies. This implies that the examination carried out in this chapter is larger in scope than that of the original version of Peluso's (2008) and Guido *et al.*'s (2008) studies. Its results, as illustrated in the next two sections, are interpretable according to an integrated approach and within a broader context, in which a consumer perspective (demand-side) is combined with that of potential players, such as producers and distributors (supply-side).

6.3. Study One: Beliefs determining acceptance of bio-fuels

6.3.1. Aims and objectives

The purpose of this study is to investigate the subjective beliefs motivating potential users to accept bio-fuels as a viable alternative to fossil fuels. By following a qualitative procedure suitable for implementing the *Theory of planned behavior* (Ajzen 1991), this study aims to find out those beliefs (behavioral, normative and control) that are mostly accessible in people's minds and capable of leading potential users to accept bio-diesel as a renewable energy source.

6.3.2. Sampling

An open-ended questionnaire was administered to a sample of 33 Italian subjects holding a direct or an indirect experience with bio-fuels or energy crops in general, as they were employed in a number of sectors directly or indirectly related to bio-fuels and agriculture. The sample almost entirely consisted of male subjects (91%), with a modal age class of 18–35 years old (60.6%). Experimental subjects held a moderate to high level of education: 45.5% of them were undergraduates; 48.5% graduates; while a marginal portion (6%) of subjects were post-graduates. This composition allows one to suppose that the selected sample was representative of the potential players' population, at least within the geographical region of interest (i.e., the Salento area in Southern Italy). These subjects could also be considered as potential bio-fuel users. A reason for this latter conclusion stems from innovation adoption literature (see Rogers 2003, for a review), which suggests that "early adopters" of an innovation (in this case, bio-diesel) tend to be aware of the new product and expert with the related technology.

6.3.3. Procedure

Following Fishbein and Ajzen (1975), an open-ended questionnaire, containing specific measures of the relevant constructs, was administered. These measures included: (*i*) one screening question asking participants to indicate at least one energy crop variety in order to assess the level of their basic knowledge in the context of reference; (*ii*) one question asking participants to indicate at least five perceived advantages and disadvantages deriving from a potential large-scale diffusion of bio-diesel and the related energy crops (*behavioral beliefs*); (*iii*) one question asking participants to indicate which subjects or groups of subjects might approve of, or disapprove of, the diffusion of this renewable energy source (*normative beliefs*); and (*iv*) one question regarding the external factors that could facilitate, or hinder, the diffusion of this technology (*control beliefs*). A further question was included in order to investigate the level of public acceptance towards the potential development of a bio-energy industry at a local level. Such a question asked participants to indicate whether and why they agree or disagree with the hypothetical construction of a biomass power station for bio-fuel production in their local area. Finally, socio-demographic data were collected (i.e., gender, class of age, education, and occupation).

6.3.4. Results of Study One

A preliminary content analysis was carried out on subjects' answers to the questionnaire. Respondents provided a satisfactory amount of information on their subjective beliefs. A content analysis was therefore performed to organize these qualitative raw data into broad answer categories, with the purpose of losing as little information as possible. A frequency analysis was performed on such data, in order to discover those beliefs that were most frequently mentioned by respondents. Based on this modal rule, such beliefs were considered to be mostly accessible in potential users' minds.

A threshold rule was adopted to discriminate between relevant and marginal beliefs. Beliefs mentioned in at least 10% of responses were considered as relevant to the present work, while those mentioned in less than 10% of cases were judged as marginal. Based on this criterion, results in Table 6.1 (*below*), regarding the behavioral beliefs associated with bio-diesel acceptance, clearly indicate the existence of a set of relevant advantages and disadvantages that subjects ascribe to bio-diesel. Economic growth was perceived to be the main advantage deriving from the diffusion of bio-diesel (23.1% of responses). Another positive consequence ascribed to this renewable energy source – and logically related to the previous one – is a growth in employment (19.7%). Besides these economic consequences, respondents identified the reduction of pollution as the third most important advantage deriving from bio-diesel (16.2%). This belief seems to be rooted in an ethical principle that could guide individuals' decision-making processes and future behaviors with regard to the adoption of this technology.

Perceived disadvantages seem to be associated with ethical principles more than advantages are. Respondents indicated a reduction of traditional organic food crops as the main negative consequence of bio-diesel diffusion (30.3%), which seems to be based upon a concern for biodiversity preservation. Next comes another serious consequence that regards a number of indirect effects on the biophysical environment (23.7%). More specifically, this answer category includes a set of ethically relevant responses pertaining to specific perceived risks.

These risks are connected with the possibility that, in the long run, bio-diesel diffusion could lead to negative phenomena such as an increase in the number of vehicles in circulation, an increase in the desertification rate of the region of reference, a reduction of forest areas, and harmful waste treatment. More than 10% of answers regarded three further disadvantages dealing with: the existence of potential inefficiencies in the production process (11.8%), which regard, for example, a low return per hectare of energy crops and an inconvenient input/output energetic ratio; a loss of local agro-food products (11.8%); and a possible over-growing of energy crops (10.5%), due to their rapid growth which means that they can spread easily in an uncontrolled manner. These latter two consequences confirm the existence of

Perceived advantages	%	*Perceived disadvantages*	%
1. Economic growth	*23.1*	*1. Reduction of traditional organic crops*	*30.3*
2. Growth of employment	*19.7*	*2. Negative indirect effects on the environment*	*23.7*
3. Pollution reduction	*16.2*	*3. Agro-production inefficiency*	*11.8*
4. Lower costs compared to traditional (fossil) fuels	8.5	*4. Loss of native local products*	*11.8*
5. Reduction in oil dependency	8.5	*5. Potential over-growth of energy crops*	*10.5*
6. Use of abandoned land	8.5	6. Multinational lobbying activity	6.6
7. Decrease in fossil fuel prices	6.8	7. Opportunistic use of incentives	5.3
8. Development of a "short" fuel supply-chain	3.4		
9. Development of an environmental mentality	2.6		
10. Increase in land returns	1.7		
11. Reduction of international political tensions	1.0		
Total	100	Total	100

Note: Relevant beliefs are reported in italics. Percentages are computed on the total answers (n) provided for perceived advantages ($n = 117$) and disadvantages ($n = 76$).

Table 6.1. Behavioral beliefs associated with energy crop and bio-diesel acceptance.

a concern for biodiversity and local food preservation among respondents, who may ascribe such risks to the use of intensive farming methods and to an excessive specialization in energy crop growing. These perceived consequences further strengthen the hypothesis that individuals are ethically concerned for environmental protection and biodiversity preservation while evaluating bio-fuel potential.

Results in Table 6.2 (*below*), regarding normative beliefs, provide clear indications about subjects or groups of subjects who could favor or hinder the diffusion of bio-diesel at a local level. Large land owners (20.4%), multinational energy firms (14.2%), the local social community (13.3%), and farmer/producer associations (10.6%) were perceived to be the most important sources of social approval and, therefore, as those subjects most capable of favoring bio-diesel diffusion in the specific social context. They might effectively serve as opinion leaders and, as such, stimulate the diffusion process among potential consumers and players in the industry. On the other hand, oil companies (27.4%), small land owners (22.6%), environmentalists (19.4%), and conservative political parties (11.3%) were perceived to be the most important sources of social disapproval and, therefore, as those subjects most capable of inhibiting bio-diesel diffusion.

Results in Table 6.3 (*below*), regarding control beliefs, provide evidence that some external factors could encourage or hinder the acceptance of this technology. Respondents perceived the presence of incentives from central authorities and banks as the main facilitator of bio-diesel acceptance (23.6% of answers). Also an increase in fossil fuel prices and a growing interest from media for new applications are considered to be important factors that could favor bio-diesel adoption (21.2% and 17.6%, respectively). Among inhibitors, the presence of unfavorable climatic conditions (20.5%), inefficiencies in the supply-chain process (19.2%), a lack of communication campaigns (13.7%), and the inertia of local economic players (e.g., entrepreneurs) are perceived as important factors capable of impeding bio-diesel diffusion.

To investigate the public acceptance of the potential development of a bioenergy industry, a frequency analysis was carried out on respondents' answers regarding their subjective predisposition towards the hypothetical construction of a biomass power station in their local area. Results showed that a vast majority of respondents declared that they agree with such an action (75.8%). Only 24.2% of respondents disagreed with the construction of such a station. A chi-square test showed that the difference between the two proportions is

Perceived sources of social approval	%	*Perceived sources of social disapproval*	%
1. Large land owners	*20.4*	*1. Oil companies*	*27.4*
2. Multinational firms in the energy industry	*14.2*	*2. Small land owners*	*22.6*
3. Local social community	*13.3*	*3. Environmentalists*	*19.4*
4. Farmer/producer associations	*10.6*	*4. Conservative political parties*	*11.3*
5. Environmentalists	9.7	5. Criminal organizations	8.1
6. Local authorities	8.0	6. Agricultural research organizations	6.5
7. Innovative entrepreneurs	8.0	7. Automotive firms	3.2
8. Local development organizations	4.4	8. Economists	1.5
9. Technical-scientific community	3.5		
10. Unemployed people	3.5		
11. Progressive political parties	3.5		
12. European Union	.9		
Total	100	Total	100

Note: Relevant beliefs are reported in italics. Percentages are computed on provided responses on perceived sources of social approval ($n = 113$) and disapproval ($n = 62$).

Table 6.2. Normative beliefs associated with energy crop and bio-diesel acceptance.

Perceived external facilitating factors	%	*Perceived external inhibiting factors*	%
1. *Incentives from central authorities and banks*	*23.6*	1. *Unfavorable climatic conditions*	*20.5*
2. *Increases in fossil fuel prices*	*21.2*	2. *Inefficiencies in the supply-chain*	*19.2*
3. *Interest from media for new applications*	*17.6*	3. *Lack of communication*	*13.7*
4. Ad hoc organizations for bio-fuel promotion	8.2	4. *Local economic players' inertia*	*13.7*
5. Economic crisis of traditional crops	7.1	5. Need to make large initial investments	8.2
6. Global warming	7.1	6. Development of more effective technologies	8.2
7. Scientific discoveries in the bio-fuel field	5.9	7. Legislative and bureaucratic constraints	8.2
8. Construction of biomass power stations	4.7	8. Corruption in politics and business	5.5
9. Effective regulation and legislation of the industry	2.3	9. Local economics crisis	1.4
10. Infrastructural investments	2.3	10. Lack of skills and competences	1.4
Total	100	Total	100

Note: Relevant beliefs are reported in italics. Percentages are computed on provided responses on perceived external facilitators ($n = 85$) and inhibitors ($n = 73$).

Table 6.3. Control beliefs associated with energy crop and bio-diesel acceptance.

significant ($\chi^2_{[1,\ 33]} = 8.785$, $p < .005$). All the justifications respondents provided to their answers were content-analyzed to identify broad response categories. A frequency analysis was therefore carried out on such data. Results showed that individuals agreeing with construction of a biomass power station justify their choice by suggesting that this action could reduce local oil dependency (40%). Further justifications provided by these respondents pertained to the possibility that the construction of such a power station could stimulate local economic growth (28%) and to the possibility that it could favor environmental protection (16%). Only 16% of these respondents provided marginal or unclear justifications to their own declarations. By contrast, the personal value of environmental protection was at the basis of the main justification provided by that 24.2% of individuals who disagreed with the hypothetical construction of a biomass power station. Results indicated that 62.5% of these respondents declared that such an investment would be environmentally unsustainable. The remaining 37.5% of subjects cited economic concerns regarding a potential lack of efficiency, efficacy, and profitability in building a biomass power station in their specific socio-geographical context.

6.4. Study Two: Beliefs determining acceptance of hydrogen transportation technologies

6.4.1. Aims and objectives

This study assesses the relative weight of a range of subjective beliefs previously identified in a preliminary study, which represented the pilot phase of the *Theory of planned behavior*'s (Ajzen 1991) implementation procedure. The pilot study was carried out through an open-ended questionnaire, similar to that developed in Study One, and administered to a convenience sample of 80 subjects in order to find out the subjective beliefs most accessible in consumers' minds. Such beliefs were thus used as items in a close-ended questionnaire within the main phase of the experimental procedure. The objective of this study is to illustrate the findings regarding the behavioral, normative, and control beliefs obtained at this stage and described in detail in Guido *et al.*'s (2008) original work.

6.4.2. Sampling

Based on the preliminary findings from the pilot phase, a close-ended questionnaire was developed and administered on a broad sample of 275 potential users of hydrogen cars. This sample consisted of 200 novice and 75 expert consumers. Expert subjects in particular included: 30 car dealers, 20 automotive managers, 15 engineers, and 10 sundry physicists and chemists. Such a distinction between novice and expert consumers was effected to verify whether any difference exists between the two groups of subjects.

6.4.3. Procedure

The close-ended questionnaire was developed from the preliminary findings as obtained from the pilot study, according to the standard procedure suggested by Fishbein and Ajzen (1975). This questionnaire included measures of the relevant constructs involved in the *Prospect method* (Caprara and Barbaranelli 2000; Caprara, Barbaranelli and Guido 2000). This method integrates the *Five-factor model* (cf. Digman 1990) and the *Theory of planned behavior* (Ajzen 1991) in a single explanatory framework for predicting how the latent dimensions describing a product's perceived image affect consumers' intention to buy that product.

The *Five-factor model* (FFM) was developed in the field of social psychology for assessing the human personality dimensions of *Agreeableness* (i.e., the subjective predisposition towards caring about others), *Conscientiousness* (i.e., the personal orientation towards goal-oriented activity), *Extroversion/Introversion* (i.e., the subjective predisposition, or aversion, towards social interaction), *Emotional stability/Neuroticism* (i.e., the subjective ability, or inability, to behave while keeping emotions and impulses under control), and *Openness to experience* (i.e., the degree of tolerance for new ideas and experiences). This theoretical framework has often applied to marketing contexts for the measurement of branded products' images, metaphorically referred to as *brand personality* (cf. Caprara, Barbaranelli and Guido 2001). The FFM was also adopted in this study for measuring the perceived image of hydrogen cars. According to this model, a list of *markers* was used for assessing such an image. They refer to specific attributes – such as *sincere*, *accurate*, *competitive*, *stable* and *modern* – which are best able to elicit the specific personality traits.

The *Prospect method* was therefore used in this study to examine how the latent dimensions of hydrogen car image, assessed through the FFM, affect potential users' purchase intention and its determinants, ascertained through the *Theory of planned behavior* (TPB). Based on this framework, questionnaire measurements included a list of 44 pairs of bipolar attributes describing the hydrogen car's perceived image. These descriptors included the 40 traditional markers of brand image, plus four product-specific attributes (i.e., *clean*, *healthy*, *respectful*, and *useful*), measured on a 7-point semantic differential scale. The questionnaire also contained a range of 32 items related to the salient behavioral, normative, and control beliefs and assessed on 7-point categorical scales, to measure the TPB's (Ajzen 1991) determinants of the potential users' intention to buy a hydrogen car.

6.4.4. Results of Study Two

A quantitative analysis was carried out on subjective beliefs which were computed by multiplying the perceived likelihood associated with their occurrence by their relative importance, according to the traditional expectancy-value approach (cf. Eagly and Chaiken 1993). A comparative analysis was, first of all, carried out on such computational products, in order to discover differences between novice and expert potential users in the relative weight (i.e., average score) ascribed to each belief. Results revealed only marginal differences, thereby showing that both novice and expert subjects held similar belief structures. This preliminary finding allowed the integration into a single dataset of the information regarding both novices' and experts' behavioral, normative and control beliefs. Results were, in turn, aggregated by following a weighted averaging procedure. In each subgroup of subjects, belief average scores were multiplied by the subgroup dimension and divided by the total sample dimension. For each subjective belief, the two weighted addends resulting from this procedure were summated to obtain an average total score measuring the relative importance of that belief for all potential users. These overall scores, measured on 49-point scales (as obtained from the expectancy-value multiplicative approach), were subsequently converted in percentage weights.

Unlike the Study One, no threshold rule was adopted in this study, as the obtained results revealed a more limited and approachable set of subjective beliefs. As regards behavioral beliefs, results in Table 6.4 indicate that health care was perceived to be the main advantage deriving from the diffusion of hydrogen as a transportation fuel (percentage weight = 21.95%).

This finding stresses how an ethical concern for healthy life could lead individuals to accept hydrogen as a new transportation technology. The relevance of personal and societal well-being as a moral principle guiding consumer perception of hydrogen cars is also confirmed by the importance level ascribed to an improved quality of life in urban areas (19.61%). However, well-being is not the only

Perceived advantages	%	*Perceived disadvantages*	%
1. Health care	*21.95*	*1. Refueling difficulties*	*24.75*
2. Environmental protection	*21.76*	*2. Worsened technical performance*	*20.20*
3. Improved quality of life in urban areas	*19.61*	*3. Hydrogen-related risks*	*20.17*
4. Independence from foreign oil suppliers	*18.46*	*4. Lowered mileage*	*18.37*
5. Fuel saving	*18.22*	*5. Limited choice possibilities*	*16.51*
Total	100	Total	100

Note: $N = 275$.

Table 6.4. Behavioral beliefs associated with hydrogen car acceptance.

principle driving consumer perception. Environmental protection is the second highest rated benefit associated with hydrogen car acceptance (21.76%). The independence from foreign oil supplies (18.46%) and the possibility of fuel saving (18.22%) are the only economic advantages perceived from hydrogen cars.

Compared to the advantages, the perceived disadvantages were based more upon technical aspects of the new technology and apparently lacking in ethical valence. They regard the possibility for potential users of encountering difficulties when refueling (percentage weight = 24.75%), of obtaining a unsatisfactory technical performance from hydrogen cars (20.20%), of running specific risks associated with the physical-chemical characteristics of hydrogen itself (20.17%), of receiving lowered mileage from hydrogen cars (18.37%), and of potentially having a limited range of car models to choose from (16.51%).

Results in Table 6.5, regarding normative beliefs, indicate that respondents perceived environmentalists as those subjects most capable

Perceived sources of social approval	%	*Perceived sources of social disapproval*	%
1. Environmentalists	*20.02*	*1. Automotive companies*	*36.89*
2. Urban area inhabitants	*19.34*	*2. Maintenance and assistance centers*	*25.07*
3. Family members	*18.92*	*3. Petrol pump owners*	*24.17*
4. Friends	*14.88*	*4. Oil supplier countries*	*13.87*
5. Local government	*14.20*		
6. Young people	*12.64*		
Total	100	Total	100

Note: $N = 275$.

Table 6.5. Normative beliefs associated with hydrogen car acceptance.

of favoring the diffusion of hydrogen cars (percentage weight = 20.02%), followed by urban area inhabitants (19.34%), family members (18.92%), friends (14.88%), local government (14.20%), and young people (12.64%). On the other hand, automotive companies were deemed to be the most important players that could disapprove of the adoption of the hydrogen technology in the transportation industry (36.89%). Respondents also perceived maintenance and assistance centers (25.07%), petrol pump owners (24.17%), and oil supplier countries (13.87%) as potential sources of opposition to hydrogen car diffusion.

Results in Table 6.6, regarding the control beliefs associated with the individual's acceptance of hydrogen cars, show that potential users perceived the introduction of incentives or the adoption of

Perceived external facilitating factors	%	*Perceived external inhibiting factors*	%
1. Improvements in urban mobility	*16.99*	*1. High prices of the hydrogen cars*	*21.30*
2. Increases in crude oil prices	*14.77*	*2. Questionable design of the hydrogen cars*	*20.35*
3. Purchase incentives	*14.60*	*3. Limited diffusion of the technology*	*19.89*
4. Management cost reduction	*14.48*	*4. Logistic inefficiencies*	*19.58*
5. Worsened environmental conditions	*13.71*	*5. Assistance inefficiencies*	*18.88*
6. Communication campaigns	*13.46*		
7. Extensive distribution of hydrogen	*11.99*		
Total	100	Total	100

Note: $N = 275$.

Table 6.6. Control beliefs associated with hydrogen car acceptance.

policies geared towards improving urban mobility as the main facilitator of the diffusion of hydrogen as a transportation fuel (percentage weight = 16.99%). They also show that respondents perceived other factors as additional facilitators of this diffusion process: an increase in crude oil prices (14.77%); the introduction of purchase incentives granted by private companies and public authorities (14.60%); a reduction of costs related to the handling of hydrogen technology (14.48%); worsening in environmental conditions (13.71%); the development of appropriate communication campaigns (13.46%); and the adoption of an extensive distribution system (11.99%).

On the other hand, the high price of hydrogen cars in the consumer market was perceived as the main factor that could inhibit acceptance of hydrogen cars (21.30%). Respondents also indicated other external factors as potential inhibitors of this acceptance process: the development of questionable designs for these cars (20.35%); a limited diffusion of the basic technology (19.89%); and the presence of inefficiencies in both logistic processes (19.58%) and assistance activities (18.88%).

Additional results reported in the original version of Guido *et al.*'s (2008) study regarded the antecedents of the novice versus expert consumers' purchase intention as determined according to the theoretical framework provided by the *Prospect method* (Caprara and Barbaranelli 2000; Caprara, Barbaranelli and Guido 2000). Although results go beyond the purpose of the present work, main findings are reported throughout the reminder of this section. In particular, it is worthwhile to illustrating the main analogies and differences between these sub-groups of subjects. As regards findings on Ajzen's (1991) model, *subjective norm*, that is, the individual perception of social pressures, and *perceived behavioral control*, namely, the perception of a personal control over external factors, were found to be significant determinants of the purchase intention for novice potential users of hydrogen cars (*subjective norm*: $\beta = .670$, $p < .001$; *perceived behavioral control*: $\beta = .215$, $p < .001$) and expert ones as well (*subjective norm*: $\beta = .617$, $p < .001$; *perceived behavioral control*: $\beta = .269$, $p < .005$).

The main difference between the two sub-groups of subjects was found in the role of *attitude*, that is, the subjective predisposition towards the purchase of a hydrogen car. *Attitude* was shown to have no significant effect on novice consumers' purchase *intention* ($p > .05$), while exerting a negative impact on experts' *intention* ($\beta = -.423$, $p < .001$). Other differences between the two sub-groups of subjects regarded the perceived image dimensions of the hydrogen car, determined according to the FFM, and their impact on purchase intention. A factor analysis was carried out on data regarding hydrogen car perceived image with the purpose of identifying the latent personality dimensions, which were properly adapted to describe the perceived image of the specific product examined. Several multiple linear regression analyses were therefore performed in order to verify whether the discovered image dimensions influence, directly or indirectly (through TPB's determinants), the potential users' behavioral *intention*. Results showed two specific dimensions, regarding *Product familiarity* (as adapted from *Agreeableness*) and its *Environmental compatibility* (as a product-specific factor), to be more able than others to describe hydrogen car image as perceived by novice consumers. The latter dimension was also found to affect purchase intention both directly ($\beta = .228$, $p < .001$) and indirectly through *perceive behavioral control* ($\beta = .154$, $p < .05$). Results showed two other dimensions, pertaining to *Product competitiveness* (as adapted from *Extroversion*) and its *Perceived originality* (as adapted from *Openness to experience*), respectively, to be prominent in experts' perception of hydrogen car image. The *Product competitiveness* factor was also found to negatively affect purchase intention through *subjective norm* ($\beta = -.251$, $p < .001$). This finding is consistent with the hypothesis that the role of social pressures in determining experts' purchase intention decreases as these subjects perceive the hydrogen technology as more competitive. On the other hand, the *Perceived originality* dimension was found to negatively affect intention through a direct path ($\beta = -.296$, $p < .001$). This result allows one to suppose that expert potential users are more reluctant to buy hydrogen cars with highly original or non-conventional designs.

6.5. General discussion

The two studies reported in this chapter provide preliminary evidence about the role of moral norms and environmental concerns in consumers' perception and acceptance of new transportation technologies, such as bio-fuels and hydrogen cars.

6.5.1. Acceptance of bio-fuels in the transportation industry

Findings from Study One show that ethical principles and moral norms are key drivers of individual evaluation processes towards controversial product categories like bio-fuels. They indicate that subjective beliefs regarding perceived advantages, social approvals, and external facilitators tend to be more heterogeneous than those concerning perceived disadvantages, social disapprovals, and external inhibitors, respectively. This result pattern means that individuals are more precise, concise, and focused on a smaller variety of arguments when evaluating the negative aspects of ethically controversial products. Results also suggest that the subjective beliefs individuals hold in their minds are rooted in their ethical principles and values. More specifically, ethical influences on judgment are stronger when individuals evaluate the negative aspects of these products, compared to the positive ones. This evaluative bias clearly emerges from results on behavioral beliefs: they show that potential bio-diesel users are more concerned about ethical issues (i.e., environmental protection and biodiversity preservation) when evaluating the disadvantages of bio-diesel diffusion than the advantages. This finding also implies that individuals are more likely to condemn unethical firms' actions than to support ethical ones. This reasoning is in line with Carrigan and Attalla's (2001) conclusions. They posit that ethically correct firms' behaviors persuade consumers to purchase their offers to a lesser extent than that to which ethical abuses encourage consumers' activism (e.g., boycott).

Ethics seems to play a marginal, indirect role in determining normative and control beliefs. In particular, no apparent link emerges from results between normative beliefs and decision makers' moral norms. It must be noted, however, that potential consumers and players in the bio-fuel industry are likely to make ethical considerations when retrieving normative beliefs from their minds. An example of this is when they mention "environmentalists" as a group of subjects capable of inhibiting energy crops and bio-diesel diffusion. Such a result may reflect an environmental orientation among respondents. This ethical principle may lead them to expect that this category of subjects will be able to impede the diffusion of energy crops (also in the light of the alleged disadvantages of bio-fuel adoption). Likewise, ethics and moral norms seem to exert a minor, indirect impact on control beliefs. Potential consumers and players in the bio-fuel industry may however assign an ethical valence to their opinions. For example, the facilitating role respondents ascribe to a possible increase in fossil fuel prices may derive from a moral norm suggesting that fuels in general should have a fair price level. This principle would lead ethical consumers and players to believe that an unjustified increase in fuel prices (i.e., a speculative pricing strategy) should favor the diffusion of alternative energy sources.

Results on individual acceptance of a biomass power station construction show that more than 70% of subjects agree with this action. This finding is interesting insofar as it contrasts with the well-known "Not-In-My-Backyard" (NIMBY) effect (Hermansson 2007). NIMBY is a widely-documented phenomenon that refers to a generalized negative attitude of residents against the decision – usually made by central or local governments or leading companies – to site an activity or a facility perceived as hazardous, or too risky, in their local area. A possible explanation of this finding may be found in the geographical context of the study. This study was carried out in the Salento area, a geographical region in Southern Italy which still lags behind other Italian regions and European countries with regard to the development of renewable energy sources (cf. Laforgia, Trevisi and De Risi 2006). For this reason, residents in this area are willing to accept the construction of a biomass station power, as they believe it would be an important source of opportunities for local economic growth and development.

6.5.2. Acceptance of hydrogen in the transportation industry

Findings from Study Two confirm that, compared to bio-fuels, hydrogen transportation technologies are less controversial from an ethical perspective. In line with Study One, results from Study Two show that potential users tend to hold a greater variety of behavioral, normative, and control beliefs with a positive valence (i.e., perceived advantages, sources of approval, and facilitators, respectively), compared to that of the same beliefs with a negative valence (i.e., perceived disadvantages, sources of disapproval, and inhibitors, respectively). This pattern of results is consistent with the hypothesis of a "pro-hydrogen bias" in potential users' perceptions and evaluations. This bias was introduced for the first time in a study by McDowall and Eames (2006: 1247), who described this phenomenon as "the individuals' tendency to perceive specific difficulties in managing hydrogen" (e.g., the difficulty of storing hydrogen) as technological challenges rather than disadvantages. A similar phenomenon could explain why positive beliefs in this survey are always equal or more numerous than negative beliefs.

Ethical principles and moral norms seem to be central when individuals evaluate positive aspects of hydrogen cars, such as their benefits to personal and public health, environmental protection, and quality of life in urban areas. This finding provides a qualitative support to the conceptual model proposed by Schulte, Hart and Van der Vorst (2004). Consistent with such a model, ethical principles and moral norms may serve as higher-order factors that determine consumers' acceptance of hydrogen transportation technologies. Technical problems regarding hydrogen storage and distribution, associated with hydrogen physical-chemical properties, seem to play a crucial role in the perception and evaluation of the negative aspects of hydrogen cars. These findings suggest that consumer ethics and moral norms could play a stimulating role in the diffusion of hydrogen cars in the mass market; but, for this to occur, automotive companies, along with researchers and technicians in energy and transport sectors, should invest money and efforts in solving technical problems related to hydrogen production and handling, as well as in developing viable alternatives to current car prototypes and models which have non-conventional designs and inaccessible price levels.

Results from the application of the *Prospect method* (Caprara, Barbaranelli and Guido 2000) show some analogies and differences between novice and expert potential users of hydrogen cars. An analogy is that, for both groups of subjects, the intention to buy a hydrogen car in the short to mid term is primarily affected by the perceived pressures towards the purchase from reference subjects or groups of subjects (e.g., environmentalists) and, secondarily, by the perception of external factors that could facilitate purchases (e.g., improvements in urban mobility). On the other hand, the different role of attitude in determining the purchase intention of novice versus expert potential users may be explained by hypothesizing the existence of different product knowledge levels for these two groups of subjects. Novice consumers are likely to lack knowledge and understanding of the effective characteristics of hydrogen cars, as well as of the consequences of their diffusion in the mass market. These subjects may consequently form an attitude that is unable to predict their purchase intention and behavior. Compared to these subjects, experts are more aware of the problems related to the adoption of hydrogen technology in civil transportation. This awareness is also likely to lead to a negative attitude, which stimulates the behavioral intention to the extent that such problems are perceived as technological challenges. A different product knowledge level in novice versus expert consumers could also explain the difference that emerged between the two groups of subjects in the perceived image dimensions of hydrogen cars. Novice consumers tend to be lacking in product knowledge. Therefore, their perception of the whole product tends to be category-based, driven by personal ethical principles and values, and focused on those aspects related to environment preservation. Expert consumers are more aware of the product's potential and its technical problems. Their perception therefore tends to be more systematic and focused on specific technical characteristics and consequences of the product.

6.5.3. Conclusions

The two studies reported in this chapter show that the success of certain new technologies launched onto the mass market does not only depend on their technical characteristics and performances. Consumers are likely to use moral principles as criteria when evaluating ethically charged innovations. These studies show that the public acceptance of new technologies in civil transportation, such as bio-fuels and hydrogen cars, depends on the degree to which such products are perceived as congruent with individuals' moral norms pertaining to ethical issues, such as environmental protection, biodiversity preservation, and societal well-being. These findings are in line with those reported in Chapters 2 and 4 (*above*), showing that *moral norms*, described as the set of personal beliefs regarding what is right or wrong, and *ethical self-identity*, defined as the extent to which individuals perceive themselves as being ethical consumers, play a determining role in the consumption of organic rather than non-organic food products.

Moreover, our results show that when an emerging technology, like hydrogen, is perceived to have a great potential to alleviate specific ethical issues at a global level, consumers tend to use different evaluative criteria depending on their level of expertise and knowledge as regards the product. Novice consumers tend to underestimate the importance of technical problems regarding with the development of product prototypes, while overestimating the relevance of potential benefits for the environment and for society at large. An opposite bias seems to emerge for expert consumers, who tend to be more focused on technical aspects and performance.

These findings have implications for companies involved in the development of ethically relevant technologies in the energy and automotive industries. At a strategic level, the adoption of ethical marketing approaches that are consistent with environmental protection, biodiversity preservation, societal well-being and other moral principles could positively affect key variables, such as new product success, change in market share (Baker and Sinkula 2005), as well as business performance in general and corporate reputation (Menon and Menon 1997). Specific strategies in such a context pertain not only to the development of "green" products or to the use of recycled raw materials in production processes, but also to the adoption of effective communication initiatives. Companies' marketing strategies should focus on specific ethical issues, which should be integrated in their strategic planning processes (cf. Banerjee 2002).

At an operational level, this reasoning implies that firms in ethically relevant businesses, in general, and in bio-fuel and hydrogen-related industries, in particular, should adopt fair pricing strategies, even at the risk of shrinking profit margins. They should invest resources in developing "green" production systems: for example, they could use modern technologies, which are based on using organic wastes rather than energy food crops as raw materials in bio-fuel production. Moreover, firms could use "short" distribution channels, in order to minimize transportation costs and other potential inefficiencies. The development of a "short" supply chain, which involves only local firms, could also exert positive effects on local economic growth and employment levels: a minimization of fuel transportation and handling costs could lead to a decrease in fuel price levels. Firms operating in the bio-fuel industry or hydrogen-related businesses should effectively integrate consumer ethical concerns for environmental protection, biodiversity preservation, and public health in their communication strategies. They could, for example, adopt cause-related marketing initiatives (cf. Guido, Prete and De Santis 2009) or develop appropriate communication campaigns aimed at informing their audiences of the environmental compatibility of their activities and products.

In conclusion, our findings suggest that, in those industries where moral principles play a driving role in consumers' perceptions and evaluations of new technologies, companies should behave as socially responsible organizations and, above all, be careful to avoid any ethical abuse when doing business.

References

Aaker, Jennifer L. 1997. Dimensions of Brand Personality. *Journal of Marketing Research*. 34/8, 347–356.

ACNielsen Homescan 2008. Retrieved February, 9th 2008 from: <http://www.acnielsen.com/products/reports/homescan>.

Aguirre, Juan A. 2001. Marketing and Consumption of Organic Products in Costa Rica. *Working Paper No. 5*. The School for Field Studies. Centre for Sustainable Development, Atenas, Costa Rica.

Aguirre, Juan A. 2007. The Farmer's Market Organic Consumer of Costa Rica. *British Food Journal*. 109/2, 145–154.

Ajzen, Icek 1991. The Theory of Planned Behavior. *Organizational Behavior and Human Decision Processes*. 50/2, 179–211.

Ajzen, Icek / Fishbein, Martin 1980. *Understanding Attitudes and Predicting Social Behavior*. Englewood Cliffs, NJ: Prentice-Hall.

Al-Khatib, Jamal A. / Vittell, Scott J. / Rawwas, Mohammed Y. A. 1997. Consumer Ethics: A Cross-Cultural Investigation. *European Journal of Marketing*. 31/11-12, 750–767.

Andersen, Esben S. / Philipsen, Kristina 1998. The Evolution of Credence Goods in Customer Markets: Exchanging "Pigs in Pokes". *Department of Business Studies*, Aalborg University, Denmark.

Antonelli, Gervasio 1996. Il Mercato dei Prodotti dell'Agricoltura Biologica: Un'Indagine in un'Ottica di Marketing. *Economia Agro-Alimentare*. 1/1, 107–146.

Armitage, Christopher J. / Conner, Mark 1999. Predictive Validity of the Theory of Planned Behavior: The Role of Questionnaire Format and Social Desirability. *Journal of Community and Applied Social Psychology*, 9/4, 261–272.

Armitage, Christopher J. / Conner, Mark 2001. Efficacy of a Minimal Intervention to Reduce Fat Intake. *Social Science & Medicine*. 52, 1517–1524.

Arvola, Anne / Vassallo, Marco / Dean, Moira / Lampila, Piritta / Saba, Anna / Lähteenmaäki, Liisa / Shepherd, Richard 2008. Predicting Intentions to Purchase Organic Food: the Role of Affective and Moral Attitudes in the Theory of Planned Behavior. *Appetite*. 50/2–3, 443–454.

Babin, Barry / Darden, William R. / Griffin, Mitch 1994, Work And/ Or Fun: Measuring Hedonic And Utilitarian Shopping Value. *Journal of Consumer Research*. 20/4, 644–655.

Baka, Jennifer / Roland-Holst, D. 2009. Food or Fuel? What European Farmers Can Contribute to Europe's Transport Energy Requirements and the Doha Round. *Energy Policy*. 37/7, 2505–2513.

Baker, Susan / Thompson, Keith E. / Engelken, Julia 2004. Mapping the Values Driving Organic Food Choice – Germany vs. the UK. *British Food Journal*. 38/8, 995–1012.

Baker, William E. / Sinkula James M. 2005. Environmental Marketing Strategy and Firm Performance: Effects on New Product Performance and Market Share. *Journal of the Academy of Marketing Science*. 33/4, 461–475.

Balat, Mustafa 2008. Potential Importance of Hydrogen as a Future Solution to Environmental and Transportation Problems. *International Journal of Hydrogen Energy*. 33/15, 4013–4029.

Bandura, Albert 1986. *Social Foundations of Thought and Action: A Social Cognitive Theory*. Englewood Cliffs, NJ: Prentice-Hall.

Bandura, Albert 1990. Mechanisms of Moral Disengagement. In Reich, Walter (ed.) *Origins of Terrorism: Psychologies, Ideologies, Theologies, States of Mind*. Cambridge: Cambridge University Press, 161–191.

Bandura, Albert 1999. Moral Disengagement in the Perpetration of Inhumanities. *Personality & Social Psychology Review*. 3/3, 193–209.

Banerjee, Subhabrata B. 2002. Corporate Environmentalism: The Construct and Its Measurement. *Journal of Business Research*. 55/3, 177–191.

Baron, Ruben M. / Kenny, David A. 1986. The Moderator-Mediator Variable Distinction in Social Psychological Research: Conceptual, Strategic, and Statistical Considerations. *Journal of Personality and Social Psychology*. 51/6, 1173–1182.

Baykara, Sema Z. 2005. Hydrogen as Fuel: A Critical Technology. *International Journal of Hydrogen Energy*. 30/5, 545–553.

Belk, Russell W. 1985. Materialism: Trait Aspects of Living in the Material World. *Journal of Consumer Research*. 12/3, 265–280.

Belk, Russell W. / Devinney, Timothy / Eckhardt, Giana 2005. Consumer Ethics Across Cultures. *Consumption, Markets and Culture*. 8/3, 275–289.

Beltramini, Richard F. 1983. Student Surrogates in Consumer Research. *Academy of Marketing Science*. 11/4, 438–443.

Bertino, Rosa M. 2006. Scuole e Aziende, Avanti con Bio. *Ristorazione Collettiva*. 4, 36–41.

Biddle, Bruce J. / Bank, Barbara J. / Slavings, Ricky L. 1987. Norms, Preferences, Identities and Retention Decisions. *Social Psychology Quarterly*. 50/4, 322–337.

Blanchette, Stephen 2008. A Hydrogen Economy and Its Impact on the World as We Know It. *Energy Policy*. 36/2, 522–530.

Bomb, Christian / McCormick, Kes / Deurwaarder, Ewout / Kåberger, Tomas 2007. Biofuels for Transport in Europe: Lessons from Germany and the UK. *Energy Policy*. 35/4, 2256–2267.

Bonti-Ankomah, Samuel / Yiridoe, Emmanuel K. 2006. Organic and Conventional Food: A Literature Review of the Economics of Consumer Perceptions and Preferences. Final Report, Organic Agriculture Center of Canada.

Bourn, Diane / Prescott, John 2002. A Comparison of the Nutritional Value, Sensory Qualities and Food Safety of Organically and Conventionally Produced Foods. *Critical Reviews in Food Science and Nutrition*. 42/1, 1–34.

Bredhal, Lone 1999. Consumers' Cognitions With Regard to Genetically Modified Foods. Results of a Qualitative Study in Four Countries. *Appetite*. 33, 343–360.

Bredhal, Lone 2001. Determinants of Consumer Attitudes and Purchase Intention with Regard to Genetically Modified Foods: Results of a Cross-National Survey. *Journal of Consumer Policy*, 24/1, 23–61.

Bredhal, Lone / Grunert, Klaus G. / Frewer, Lynn J. 1998. Consumer Attitudes and Decision-Making with Regard to Genetically Engineered Food Products – A Review of the Literature and a Presentation of Models for Future Research. *Journal of Consumer Policy*. 21/3, 251–277.

Buchan, Howard F. 2005. Ethical Decision Making in the Public Accounting Profession: An Extension of Ajzen's Theory of Planned Behavior. *Journal of Business Ethics*. 61/2, 165–181.
Buzby, Jean C. / Skees, Jerry R. 1994. Consumers Want Reduced Exposure to Pesticides in Food. *Food Review*. 17/2, 19–22.
Byrne, Patrick J. / Bacon, Richard / Toensmeyer, Ulrich C. 1994. Pesticide Residue Concerns and Shopping Location Likelihood. *Agribusiness*. 10/6, 491–501.
Canavari, Maurizio / Bazzani, Guido M. / Spadoni, Roberta / Regazzi, Domenico 2002. Food Safety and Organic Fruit Demand in Italy: A Survey. *British Food Journal*. 104/3-5, 220–232.
Canavari, Maurizio / Nayga, Rodolfo M. 2009. On Consumers' Willingness to Purchase Nutritionally Enhanced Genetically Modified Food. *Applied Economics*. 41/1, 125–137.
Caprara, Gian Vittorio / Barbaranelli, Claudio 1996. La Danza degli Aggettivi. *Micro e Macro Marketing*. 5, 7–21.
Caprara, Gian Vittorio / Barbaranelli, Claudio 2000. *Capi di Governo, Telefonini, Bagni Schiuma: Determinanti Personali dei Comportamenti di Voto e di Acquisto*. Milan: Cortina Editore.
Caprara, Gian Vittorio / Barbaranelli, Claudio / Guido, Gianluigi 1998. Personality as Metaphor: Extension of the Psycholexical Hypothesis and the Five Factor Model to Brand and Product Personality Description. In Englis, Basil / Olafsson, Anna (eds) *European Advances in Consumer Research*. Provo, UT: Association for Consumer Research. 3, 61–69.
Caprara, Gian Vittorio / Barbaranelli, Claudio / Guido, Gianluigi 2000. The Prospect Model: How to Analyze and Forecast the Influence of Brand and Product Personality on Determinants of Purchase Intention. *Testing Psicometria Metodologia*. 7/3-4, 113–128.
Caprara, Gian Vittorio / Barbaranelli, Claudio / Guido, Gianluigi 2001. Brand Personality: How to Make the Metaphor Fit? *Journal of Economic Psychology*. 22/3, 377–395.
Caprara, Gian Vittorio / Barbaranelli, Claudio / Vicino, Sabrina / Bandura, Albert 1996. La Misura del Disimpegno Morale. *Rassegna di Psicologia*. 13, 93–105.
Caprara, Gian Vittorio / Barbaranelli, Claudio / Pastorelli, Concetta / Iafrate, Claudia / Beretta, Mara / Steca, Patrizia / Bandura, Al-

bert 2006. La Misura del Disimpegno Morale nel Contesto delle Trasgressioni dell'Agire Quotidiano. *Giornale Italiano di Psicologia*. 33/1, 83–106.
Carboni, Roberto 2002. Le Dinamiche del Biologico. *Economia Agro-Alimentare*. 7/2, 121–135.
Carey, Lindsey / Shaw, Deidre / Shiu, Edward 2008. The Impact of Ethical Concerns on Family Consumer Decision-Making. *International Journal of Consumer Studies*. 32/5, 553–560.
Carrigan, Marylyn / Attalla, Ahmad 2001. The Myth of the Ethical Consumer – Do Ethics Matter in Purchase Behavior? *Journal of Consumer Marketing*. 18/7, 560–577.
Caruana, Robert 2007. A Sociological Perspective of Consumption Morality. *Journal of Consumer Behaviour*. 6/5, 287–304.
Cava, Veronica E. 2000. Moral Disengagement and Academic Cheating: The Role of Individual Difference and Situational Variables. *Dissertation Abstracts International Section A: Humanities and Social Sciences*. 61/2-A, 492.
Charng, Hong-Wen / Piliavin, Jane A. / Callero, Peter L. 1988. Role Identity and Reasoned Action in the Prediction of Repeated Behavior. *Social Psychology Quarterly*. 51/4, 303–317.
Chatzidakis, Andreas / Hibbert, Sally / Smith, Andrew 2006. "Ethically Concerned, Yet Unethically Behaved": Towards An Updated Understanding of Consumer's (Un)Ethical Making. In Pechmann, Connie / Price, Linda (eds) *Advances in Consumer Research*. Duluth, MN: Association for Consumer Research. 33, 693–698.
Chen, Mei-Fang 2007. Consumer Attitudes and Purchase Intention in Relation to Organic Foods in Taiwan: Moderating Effects of Food-Related Personality Traits. *Food Quality and Preference*. 18/7, 1008–1021.
Cherry, Robert S. 2004. A Hydrogen Utopia? *International Journal of Hydrogen Energy*. 29/2, 125–129.
Chinnici, Gaetano / D'Amico, Mario / Pecorino, Biagio 2002. A Multivariate Statistical Analysis on the Consumers of Organic Products. *British Food Journal*. 104/3-5, 187–199.
Cicia, Gianni / Del Giudice, Teresa / Scarpa, Riccardo 2002. Consumers' Perception of Quality in Organic Food: A Random Utility

Model Under Preference Heterogeneity and Choice Correlation from Rank-Orderings. *British Food Journal*. 104/3-5, 200–213.
Clarke, Geraldine 2006. The Ethical Consumer. *Journal of Consumer Behaviour*. 5/1, 95–96.
Codeluppi, Vanni (ed.) 2003. *La Sfida della Pubblicità*. Milan: Franco Angeli.
Colacchio, Fabio / Mastrangelo, Dania / Zarantonello, Lia 2008. I Vissuti del Cibo: Una Ricerca Esplorativa. *Mercati e Competitività*. 4/3, 41–69.
Conner, Mark / Armitage, Christopher 1998. Extending the Theory of Planned Behavior: A Review and Avenues for Further Research. *Journal of Applied Social Psychology*. 28/15, 1429–1464.
Conner, Mark / Smith, Neil / McMillan, Brian 2003. Examining Normative Pressure in the Theory of Planned Behaviour: Impact of Gender and Passengers on Intentions to Break the Speed Limit. *Current Psychology*. 22/3, 252–263.
Conner, Mark / Lawton, Rebecca / Parker, Dianne / Chorlton, Kathryn / Manstead, Antony S.R. / Stradling, Stephen 2007. Application of the Theory of Planned Behaviour to the Prediction of Objectively Assessed Breaking of Posted Speed Limits. *British Journal of Psychology*. 98/3, 429–453.
Connolly, John / Shaw, Deirdre 2006. Identifying Fair Trade in Consumption Choice. *Journal of Strategic Marketing*. 14/4, 353–368.
Cooper-Martin, Elizabeth / Holbrook, Morris B. 1993. Ethical Consumption Experiences and Ethical Space. In McAlister, Leigh / Rothschild Michael L. (eds.) *Advances in Consumer Research*. Provo, UT: Association for Consumer Research. 20, 113–118.
Crane, Andrew / Matten, Dirk 2004. *Business Ethics: A European Perspective*. Oxford: Oxford University Press.
Darby, Michael R. / Karni, Edi 1973. Free Competition and the Optimal Amount of Fraud. *Journal of Law and Economics*. 16, 67–88.
D'Astous, Alain / Lévesque, Mélanie 2003. A Scale for Measuring Store Personality. *Psychology & Marketing*. 20/5, 455–469.
Datamonitor 2005. *Ethical Consumerism: France Sets the Standard*, Retrieved April, 20th 2008 from: <http://www.datamonitor.com /industries/news/article/?pid=12B06A73-7423-493A-90E1-1DB 65A461D5A&type=CommentWire>.

Datamonitor 2007. *Organic Food: Global Industry Guide*. Retrieved April, 20th 2008 from: <http://www.datamonitor.com>.
Davies, Anne / Titterington, Albert J. / Cochrane, Clive 1995. Who Buys Organic Food? A Profile of the Purchasers of Organic Food in Northern Ireland. *British Food Journal*. 97/10, 17–23.
Dean, Moira / Arvola, Anne / Vassallo, Marco / Lahteenmaki, Liisa / Raats, Monique M. / Saba, Anna / Shepherd, Richard 2006. Comparison of Elicitation Methods for Moral and Affective Beliefs in the Theory of Planned Behavior. *Appetite*. 47/2, 244–252.
De Jonge, Janneke / Frewer, Lynn J. / Van Trijp, Hans / Renes, Reint J. / de Wit, Willem / Timmers, Joke 2004. Monitoring Consumer Confidence in Consumer Safety: an Exploratory Study. *British Food Journal*. 106, 10/11, 837–849.
Demirbas, Ayhan 2007a. Importance of Biodiesel as Transportation Fuel. *Energy Policy*. 35/9, 4661–4670.
Demirbas, Ayhan 2007b. Progress and Recent Trends in Biofuels. *Progress in Energy and Combustion Science*. 33/1, 1–18.
Dennison, Catherine M. / Shepherd, Richard 1995. Adolescent Food Choice: An Application of the Theory of Planned Behaviour. *Journal of Human Nutrition and Dietetics*. 8, 9–23.
De Ruvo, Enrico 2007. *Il Mercato dei Prodotti Biologici: Tendenze Generali nelle Principali Filiere*. Rome: ISMEA.
Devos, Yann / Maeseele, Pieter / Reheul, Dirk / Van Speybroeck, Linda / De Waele, Danny 2007. Ethics in the Societal Debate on Genetically Modified Organisms: a (Re)quest for Sense and Sensibility. *Journal of Agricultural and Environmental Ethics*. 21, 29–61.
Digman, John M. 1990. Personality Structure: Emergence of the Five-Factor Model. *Annual Review of Psychology*. 41/1, 417–440.
Dinse, Gundi 1999. Wasserstofffahrzeuge und ihr Funktionsraum – Eine Analyse der technischen, politisch-rechtlichen und sozialen Dimension. Working Paper. University of Berlin, Germany.
Dinse, Gundi 2000. Akzeptanz von wasserstoffbetriebenen Fahrzeugen – Eine Studie über die Verwendung eines neuen und ungewohnten Kraftstoffs. Working Paper. Institut für Mobilitätsforschung.
Dobni, Dawn and Zinkhan, George M. 1990. In Search of Brand Image: A Foundation Analysis. In Goldberg, Martin E. *et al.* (eds),

Advances in Consumer Research, Provo, UT: Association for Consumer Research. 17, 110–119.
Doran, Caroline J. 2009. The Role of Personal Values in Fair Trade Consumption. *Journal of Business Ethics*. 84, 549–563.
Dreezens, Ellen / Martijn, Carolien / Tenbült, Petra / Kok, Gerjo / De Vries, Nanne K. 2005. Food and Values: an Examination of Values underlying Attitudes toward Genetically Modified and Organically Grown Food Products. *Appetite*. 44/1, 115–122.
Dunn, Seth 2002. Hydrogen Futures: Toward a Sustainable Energy System. *International Journal of Hydrogen Energy*. 27/3, 235–264.
Eagly, Alice H. / Chaiken, Shelly 1993. *The Psychology of Attitudes*. San Diego, CA: Harcourt Brace Jovanovich.
Eagly, Alice H. / Chaiken, Shelly 2007. The Advantage of an Inclusive Definition of Attitude. *Social Cognition*. 25/5, 582–602.
Eartmans, Audrey / Victoir, An / Vansant, Greet / Van den Bergh, Omer 2005. Food-Related Personality Traits, Food Choice Motives and Food Intake: Mediator and Moderator Relationships. *Food Quality and Preference*. 16, 714–726.
Edwards, Peter P. / Kuznetsov, Vladimir L. / David, William I. F. / Brandon, Nigel P. 2008. Hydrogen and Fuel Cells: Towards a Sustainable Energy Future. *Energy Policy*. 36/12, 4356–4362.
Elliott, Dave 2000. Renewable Energy and Sustainable Futures. *Futures*. 32/3-4, 261–274.
Eurobarometer 2006. Risk Issues, *Special Eurobarometer 238*. Brussels: European Commission. Retrieved November 29, 2008 from: <http://ec.europa.eu/public_opinion/archives/ebs/ebs_238_en.pdf>.
European Commission 2006. An EU Strategy for Biofuels. Retrieved May, 18th 2009 from: <http://ec.europa.eu/agriculture/biomass/biofuel/com2006_34_en.pdf>.
Farrell, Alexander E. / Keith David W. / Corbett James J. 2003. A Strategy for Introducing Hydrogen into Transportation. *Energy Policy*. 31/13, 1357–1367.
Feather, Norman T. 1992. Values, Valences, Expectations, and Actions. *Journal of Social Issues*. 48/2,109–124.
Fiddes, N. 1989. Meat: A Natural Symbol. New York: Routledge.
Field, Andy P. 2005. *Discovering Statistics Using SPSS*. London: Sage.

Finch, James E. 2005. The Impact of Personal Consumption Values and Beliefs on Organic Food Purchase Behavior. *Journal of Food Products Marketing*. 11/4, 63–76.

Finucane, Melissa / Slovic, Paul / Mertz, Chris K. / Flynn, James / Satterfield, Theresa A. 2000. Gender, Race, and Perceived Risk: the "White Male" Effect. *Health, Risk & Society*. 2/2, 159–172.

Fishbein, Martin / Ajzen, Icek 1975. *Belief, Attitude, Intention, and Behavior: An Introduction to Theory and Research*. Reading, MA: Addison-Wesley.

Fisher, Robert J. 1993. Social Desirability Bias and the Validity of Indirect Questioning. *Journal of Consumer Research*. 20/2, 303–315.

Follows, B. Scott / Jobber, David 2000. Environmentally Responsible Purchase Behaviour: A Test of a Consumer Model. *European Journal of Marketing*. 34/5-6, 723–746.

Food Agriculture Organization, 2008. Bioenergy, Food Security and Sustainability – Towards an International Framework. High-Level Conference on World Food Security: The Challenges of Climate Change and Bioenergy. Rome. Retrieved May, 18th 2009 from: <http://www.fao.org/fileadmin/user_upload/foodclimate/HLCdocs/HLC08-Rep-E.pdf>.

Fotopoulos, Christos / Chryssochoidis, George 2000. Factors Affecting the Decision to Purchase Organic Food. *Journal of Euromarketing*. 9/3, 45–66.

Fotopoulos, Christos / Krystallis, Athanasios 2002. Organic Product Avoidance: Reasons for Rejection and Potential Buyers' Identification in a Countryside Survey. *British Food Journal*. 104/3-5, 233–260.

Freestone, Oliver M. / McGoldrick, Peter J. 2008. Motivations of the Ethical Consumer. *Journal of Business Ethics*. 79, 445–467.

Frewer, Lynn / Lassen, Jesper / Kettliz, Beate / Scholderer, Joachim / Beekman, Volkert / Berdal, Knut G. 2004. Societal Aspects of Genetically Modified Foods. *Food and Chemical Toxicology*. 42, 1181–1193.

Frondel, Manuel / Peters, Jörg 2007. Biodiesel: A New Oildorado? *Energy Policy*. 35/3, 1675–1684.

Fukukawa, Kyoko 2003. A Theoretical Review of Business and Consumer Ethics Research: Normative and Descriptive Approaches. *The Marketing Review*. 3, 381–401.

Gaskell, George / Allum, Nick / Wagner, Wolfang / Kronberger, Nicole / Torgersen, Helge / Hampel, Juergen / Bardes, Julie 2004. GM Foods and the Misperception of Risk Perception. *Risk Analysis*. 24/1, 185–194.

Geganov, Encho N. / Dilova, Margarita L. / Petkova, Kristina G. / Paspalanova, Elena P. 1996. Culture-Specific Approach to the Study of Individualism/Collectivism. *European Journal of Social Psychology*. 26, 277–297.

Giannakas, Kostantinos 2002. Information Asymmetries and Consumption Decisions in Organic Food Product Markets. *Canadian Journal of Agricultural Economics*. 50, 35–50.

Gifford, Katie / Bernard, John C. 2006. Influencing Consumer Purchase Likelihood of Organic Food. *International Journal of Consumer Studies*. 30/2, 155–163.

Gokcol, Cihan / Dursun, Bahtiyar / Alboyaci, Bora / Sunan Erkan 2009. Importance of Biomass Energy as Alternative to Other Sources in Turkey. *Energy Policy*. 37/2, 424–431.

Goldberg, Lewis R. 1993. The Structure of Phenotypic Personality Traits. *American Psychologist*. 48/1, 23–34.

Goldman, Barbara J. / Clancy, Katherine L. 1991. A Survey of Organic Produce Purchases and Related Attitudes of Food Cooperative Shoppers. *American Journal of Alternative Agriculture*. 6/2, 89–92.

Goldsmith, Ronald E. / Frieden, Jon / Henderson, Kenneth V. 1997. The Impact of Social Values on Food-Related Attitudes. *British Food Journal*. 99/9, 352–357.

Govindasamy, Ramu / Italia, John 1997. Consumer Response to Integrated Pest Management and Organic Agriculture: An Econometric Analysis. *New Jersey Agricultural Experiment Station*, Cook College, Rutgers: The State University of New Jersey.

Granberg, Donald. / Holmberg, Soren 1990. The Intention-Behavior Relationship Among U.S. and Swedish Voters. *Social Psychology Quarterly*. 53/1, 44–54.

Groff, Andrew J. / Kreidor, Craig R. / Toensmeyer, Ulrich C. 1993. Analysis of the Delaware Market for Organically Grown Products. *Journal of Food Distribution Research.* 24/1, 118–125.

Grunert, Klaus G. / Bech-Larsen, Tino / Lähteenmäki, Liisa / Ueland, Øydis / Åström, Annika 2004. Attitudes towards the Use of GMOs in Food Production and their Impact on Buying Intention: the Role of Positive Sensory Experience. *Agribusiness.* 20/1, 95–106.

Grunert, Suzanne C. / Juhl, Hans J. 1995. Values, Environmental Attitudes, and Buying of Organic Foods. *Journal of Economic Psychology.* 16/1, 39–62.

Guido, Gianluigi 1998. The Dichotic Theory of Salience: A Framework for Assessing Attention and Memory. *European Advances in Consumer Research.* In Englis, Basil / Olafsson, Anna (eds) *European Advances in Consumer Research.* Provo, UT: Association for Consumer Research. 3, 114–119.

Guido, Gianluigi 2001. *The Salience of Marketing Stimuli. An Incongruity-Salience Hypothesis on Consumer Awareness.* Boston, MA: Kluwer Academic Publishers.

Guido, Gianluigi 2005. La Dimensione dei Rapporti con i Consumatori nelle Imprese Orientate al Marketing Competitivo. In Pilotti, Luciano (ed.) *Strategie dell'Impresa.* Rome: Carocci Editore, 113–155.

Guido, Gianluigi / Peluso, Alessandro M. / Prete, M. Irene / Quarchioni, Juri 2008. The Market for Hydrogen Cars: Non-Expert and Expert Consumers' Product Images and Determinants of Purchase Intention. *The World of New Economy.* 2, 41–47.

Guido, Gianluigi / Prete, M. Irene / De Santis, Danila 2009. The Role of Cause-Related Marketing in "Co-opetition with Consumers" Strategies. *Advances in Management.* 2/1, 7–16.

Gupta, Ram B. (ed.) 2008. *Hydrogen Fuel: Production, Transport, and Storage.* Boca Raton, FL: Taylor and Francis.

Hansen, Lars G. 2001. Modelling Demand for Organic Products – Implications for the Questionnaire. *Working Paper No 4.* AKF, Danish Institute of Local Government Studies. Retrieved May, 7th 2009 from: <http://www.akf.dk/organicfoods/papers/wp4-lgh.pdf>.

Harper, Gemma C. / Mikatouni, Aikaterini 2002. Consumer Perception of Organic Food Production and Farm Animal Welfare. *British Food Journal.* 104/3, 287–299.

Harris, Michael J. 1997. Consumers pay a premium for organic baby foods. *Food Review.* May-August: 13–16.

Harrison, Rob / Newholm, Terry / Shaw, Deirdre 2005. *The Ethical Consumer.* London: Sage.

Haukenes, Anne 2004. Perceived Health Risks and Perceptions of Expert Consensus in Modern Food Society. *Journal of Risk Research.* 7, 759–774.

Hermansson, Hélène 2007. The Ethics of NIMBY Conflicts. *Ethics Theory and Moral Practice.* 10/1, 23–34.

Hickson, Allister / Phillips, Al / Morales, Gene 2007. Public Perception to a Hydrogen Hybrid Internal Combustion Engine Transit Bus Demonstration and Hydrogen Fuel. *Energy Policy.* 35/4, 2249–2255.

Hill, Helene / Lynchehaun, Fidelma 2002. Organic Milk: Attitudes and Consumption Patterns. *British Food Journal.* 104/7, 526–542.

Hofstede, Geert 1983. Dimensions of National Cultures in Fifty Countries and Three Regions. In Deregowski, Jan B. / Dziurawiec, Susanne / Annis, Robert C. (eds) *Explications in Cross-Cultural Psychology.* Lisse, Holland: Swets & Zeitlinger, 335–355.

Hofstede, Geert 1984. Cultural Dimensions in Management and Planning. *Asia Pacific Journal of Management.* 1/2, 81–99.

Hofstede, Geert 2001. *Culture's Consequences: Comparing Values, Behaviors, Institutions and Organizations across Nations.* Thousand Oaks, CA: Sage.

Hofstede, Geert 2009. Geert Hofstede – Cultural Dimensions. Retrieved April, 5th 2009 from: <http://www.geert-hofstede.com>.

Homer, Pamela M. / Kahle, Lynn R. 1988. A Structural Equation Test of the Value-Attitude-Behavior Hierarchy. *Journal of Personality and Social Psychology.* 54/4, 638–646.

Honkanen, Pirjo / Verplanken, Bas / Olsen, Svein O. 2006. Ethical Values and Motives Driving Organic Food Choice. *Journal of Consumer Behaviour.* 5/5, 420–430.

Honkanen, Pirjo / Verplanken, Bas 2004. Understanding Attitudes towards Genetically Modified Food. *Journal of Consumer Policy*. 27, 401–420.

Hossain, Ferdaus / Onyango, Benjamin 2004. Product Attributes and Consumer Acceptance of Nutritionally Enhanced Genetically Modified Foods. *International Journal of Consumer Studies*. 28/3, 255–267.

Hughner, Renée S. / McDonagh, Pierre E. / Prothero, Andrea / Shultz II, Clifford J. / Stanton, Julie 2007. Who are Organic Food Consumers? A Compilation and Review of why People Purchase Organic Food. *Journal of Consumer Behaviour*. 6/2–3, 94–110.

Hutchins, Richard K. / Greenhalgh, L.A. 1997. Organic Confusion: Sustaining Competitive Advantage. *British Food Journal*. 99/9, 336–338.

International Federation of Organic Agriculture Movements 2005. Retrieved February, 9th 2008 from: <http://www.ifoam.org/about_ifoam/principles/index.html>.

International Federation of Organic Agriculture Movements 2009. *Definition of Organic Agriculture* Retrieved March, 12th 2009 from: <http://www.ifoam.org/growing_organic/definitions/doa/index.html>.

"Is there such a Thing as an Ethical Consumer?" 2003. *Strategic Direction*. 19/6, 28–30.

Jaccard, James / Blanton, Hart 2005. The Origins and Structure of Behavior: Conceptualizing Behavior in Attitude Research. In Albarracin, Dolores / Johnson, Blair T. / Zanna, Mark P. (eds) *The Handbook of Attitudes*. Mahwah, NJ: Lawrence Erlbaum, 125–172.

James, Clive 2007. Global Status of Commercialized Biotech/GM crops: 2007, ISAAA Brief No. 37. Ithaca, NY: ISAAA. Retrieved November 29th, 2008 from: <http://www.isaaa.org>.

Jolly, Desmond A. 1991. Differences between Buyers and Non-Buyers of Organic Produce and Willingness to Pay Organic Price Premiums. *Journal of Agribusiness*. 9/1, 97-111.

Johnston, Brenda / Mayo, Michael C. / Khare, Anshuman 2005. Hydrogen: The Energy for the 21st Century. *Technovation*. 25/6, 569–585.

Karp, David G. 1996. Values and their Effect on Pro-Environmental Behavior. *Environment and Behavior*. 28/1, 111–133.

Koh, Lian P. 2007. Potential Habitat and Biodiversity Losses from Intensified Biodiesel Feedstock Production. *Conservation Biology*. 21/5, 1373–1375.

Kolodinsky, Jane / DeSisto, Thomas P. / Labrecque, Joanne 2003. Understanding the Factors Related to Concerns over Generically Engineered Food Products: Are National Differences Real? *International Journal of Consumer Studies*. 27/4, 266–276.

Laforgia, Domenico / Trevisi, Antonio S. / De Risi, Arturo 2006. Potenzialità delle Risorse Rinnovabili nel Mezzogiorno. *Energia*. 27/3, 46–58.

Langdridge, Darren / Sheeran, Paschal / Connolly, Kevin J. 2007. Analyzing Additional Variables in the Theory of Reasoned Action. *Journal of Applied Social Psychology*. 37/8, 1884–1913.

Laure, Bruno / West, Gale E. / Gendron, Carole / Lambert, Rémy 2004. Consumer Response to Functional Foods Produced by Conventional, Organic, or Genetic Manipulation. *Agribusiness*. 20/2, 155–166.

Lea, Emma / Worsley, Tony 2005. Australians' Organic Food Beliefs, Demographics and Values. *British Food Journal*. 107/11, 855–869.

Leatherhead Food International 2007. *The Ethical Consumer – A Global Perspective*. Leatherhead, UK: Leatherhead Food International.

Lockie, Stewart / Lyons, Kristen / Lawrence, Geoffrey / Mummery, Kerry 2002. Eating "Green": Motivations Behind Organic Food Consumption in Australia. *Sociologia Ruralis*. 42/1, 23–40.

Lofordos, George N. / Dennis, June 2008. Consumers' Intent: In the Organic Food Market. *Journal of Food Products Markets*. 14/2, 17–38.

Ludwig-Bölkow-Systemtechnik 1998. The Acceptance of Hydrogen Technologies. Executive Summary. Retrieved May, 18th 2009 from: <http://www.hydrogen.org/accepth2/execsumm .html>.

MacKinnon, David P. / Fairchild, Amanda J. / Fritz, Matthew S. 2007. Mediation Analysis. *Annual Review of Psychology*. 58, 593–614.

Magkos, Faidon / Arvaniti, Fotini / Zampelas, Antonis 2003. Organic Food: Nutritious Food or Food for Thought? A Review of the Evidence. *International Journal of Food Sciences and Nutrition.* 54/5, 357–371.
Magkos, Faidon / Arvaniti, Fotini / Zampelas, Antonis 2006. Organic Food: Buying More Safety or just Peace of Mind? A Critical Review of the Literature. *Critical Reviews in Food Science and Nutrition.* 46, 23–56.
Magnusson, Maria K. / Arvola, Anne / Koivisto Hursti, Ulla K. / Åberg, Lars / Sjödén, Per O. 2001. Attitudes towards Organic Foods among Swedish Consumers. *British Food Journal.* 103/3, 209–227.
Magnusson, Maria K. / Arvola, Anne / Koivisto Hursti, Ulla K. / Åberg, Lars / Sjödén, Per O. 2003. Choice of Organic Foods is Related to Perceived Consequences for Human Health and to Environmentally Friendly Behavior. *Appetite.* 40, 109–117.
Maio, Gregory R. / Olson, James M. / Allen, Lindsay / Bernard, Mark M. 2001. Addressing Discrepancies Between Values and Behavior: the Motivating Effect of Reasons. *Journal of Experimental Social Psychology.* 37, 104–117.
Makatouni, Aikaterini 2002. What Motivates Consumers to Buy Organic Food in the UK? Results from a Qualitative Study. *British Food Journal.* 104/3-5, 345–352.
Manget, Joe / Roche, Catherine / Münnich, Felix 2009. Capturing the Green Advantage for Consumer Companies. *Boston Consulting Group Research Report.* Retrieved April, 12th 2009 from: <http://www.bcg.com/impact_expertise/publications/files/Captuing_Green_Advantage_Consumer_Companies_Jan_2009.pdf>.
Manstead, Antony S.R. 1999. The Role of Moral Norm in the Attitude-Behavior Relationship. In Terry, Deborah J. / Hogg Michael A. (eds) *Attitudes, Behavior, and Social Context: The Role of Norms and Group Membership.* Mahwah, NJ: Lawrence Erlbaum.
Mathews, John A. / Goldsztein, Hugo 2009. Capturing Latecomer Advantages in the Adoption of Biofuels: The Case of Argentica. *Energy Policy.* 37/1, 326–337.
McDowall, William / Eames, Malcolm 2006. Forecasts, Scenarios, Visions, Backcasts and Roadmaps to the Hydrogen Economy:

A Review of the Hydrogen Futures Literature. *Energy Policy*. 34/11, 1236–1250.

McEachern, Morven G. / McClean, Paula 2002. Organic Purchasing Motivations and Attitudes: Are They Ethical? *International Journal of Consumer Studies*. 26/2, 85–92.

McKay, Betsy 2007. When Buying Organic Makes Sense and When It Doesn't. *Wall Street Journal – Eastern Edition*. 249/12, D1–D2.

McMillan, Brian / Conner, Mark 2003. Applying an Extended Version of the Theory of Planned Behavior to Illicit Drug Use Among Students. *Journal of Applied Social Psychology*. 33/8, 1662–1683.

Menon, Ajay / Menon Anil 1997. Enviropreneurial Marketing Strategy: The Emergence of Corporate Environmentalism as Market Strategy. *Journal of Marketing*. 61/1, 51–67.

Michaelidou, Nina / Hassan, Louise M. 2008. The Role of Health Consciousness, Food Safety Concern and Ethical Identity on Attitudes and Intentions Towards Organic Food. *International Journal of Consumer Studies*. 32/2, 163–170.

Minton, Ann P. / Rose, Randall L. 1997. The Effects of Environmental Concern on Environmentally Friendly Consumer Behavior: An Exploratory Study. *Journal of Business Research*. 40/1, 37–48.

Morrone, Michele / Stuart Ben J. / McHenry, Izaak / Buckley Geoffrey L. 2009. The Challenges of Biofuels from the Perspectives of Small-Scale Producers in Ohio. *Energy Policy*. 37/2, 522–530.

Muncy, James A. / Eastman, Jacqueline K. 1998. Materialism and Consumer Ethics: An Exploratory Study. *Journal of Business Ethics*. 17/2, 137–145.

Muncy, James A. / Vitell, Scott J. 1992. Consumer Ethics: An Investigation of the Ethical Beliefs of the Final Consumer. *Journal of Business Research*. 24/4, 297–311.

Naspetti, Simona / Zanoli, Raffaele 2006. Organic Food Quality and Safety Perception throughout Europe. Paper presented at the 98th EAAE Seminar: *Marketing Dynamics within the Global Trading System: New Perspectives*. Chania, CR, June, 29–July, 2. Retrieved November 27th, 2008 from: <http://209.85.129.132/search?q=cache:NeLUQK8mDi8J:ageconsearch.umn.edu/bitstream/10086/1/sp06na01.pdf+naspetti+zanoli+%22organic+food+quality%22&hl=it&ct=clnk&cd=3&gl=it>.

Newholm, Terry / Shaw, Deirdre 2007. Studying the Ethical Consumer: A Review of Research. *Journal of Consumer Behaviour*. 6/5, 253–270.

Ng, Siew I. / Lee, Julie A. / Soutar, Geoffrey N. 2007. Are Hofstede's and Schwartz's Value Frameworks Congruent? *International Marketing Review*. 24/2, 164–180.

O'Connor, Rory C. / Armitage, Christopher J. 2003. Theory of Planned Behaviour and Parasuicide: An Exploratory Study. *Current Psychology*. 22/3, 196–205.

O'Donovan, Pym / McCarthy, Mary 2002. Irish Consumer Preference for Organic Meat. *British Food Journal*. 104/3-5, 353–370.

Organic Trade Association 2007. *The Organic Trade Association's 2007 Manufacturer Survey*. Retrieved February, 9th 2008 from: <http: //www.ota.com/pics/documents/2007ExecutiveSummary.pdf>.

Organic Trade Association 2008. Retrieved February, 9th 2008 from: <http://www.ota.com/definition/quickoverview.html>.

Organic Trade Association 2009. Retrieved February 9th from: <http://www.ota.com/definition/quickoverview.html>.

Padel, Susanne / Foster, Carolyn 2005. Exploring the Gap Between Attitudes and Behaviour: Understanding Why Consumers Buy or Do Not Buy Organic Food. *British Food Journal*. 107/8, 606–625.

Pahl, Greg 2005. *Biodiesel: Growing a New Energy Economy*. Chelsea: Chelsea Green.

Parker, Dianne / Manstead, Antony S.R. / Stradling, Stephen G. 1995. Extending the Theory of Planned Behavior: The Role of Personal Norm. *British Journal of Social Psychology*. 34/2, 127–137.

Parkins, Wendy / Craig, Geoffrey 2006. *Slow Living*. Oxford: Berg.

Peluso, Alessandro M. 2008. Biodiesel e Centrali a Biomassa nel Salento: Uno Studio Esplorativo sulle Credenze dei Potenziali Operatori del Settore. *Economia, Azienda e Sviluppo*. 4/3, 83–102.

Perrez, Franz X. 2000. Taking Consumers Seriously: the Swiss Regulatory Approach to Genetically Modified Food. *N.Y.U. Environmental Law Journal*. 8, 585–604.

Pimentel, David / Patzek, Tad W. 2005. Ethanol Production Using Corn, Switchgrass, and Wood; Biodiesel Production Using Soybean and Sunflower. *Natural Resources Research*. 14/1, 65–76.

Plummer, Joseph T. 2000. How Personality Makes a Difference. *Journal of Advertising Research*. 40/6, 79–83.

Poiesz, Theo B.C. 1989. The Image Concept: Its Place in Consumer Psychology. *Journal of Economic Psychology*. 10, 457–472.

Polonsky, Michael J. / Brito, Pedro Q. / Higgs-Kleyn, Nicola 2001. Consumer Ethics in the European Union: A Comparison of Northern and Southern Views. *Journal of Business Ethics*. 31/2, 117–130.

Portinga, Wouter / Pidgeon, Nick F. 2006. Exploring the Structure of Attitudes towards Genetically Modified Food. *Risk Analysis*. 26/6, 1707–1719.

Post, James E. 1985. Assessing the Nestlé Boycott: Corporate Accountability and Human Rights. *California Management Review*. 27/2, 113–131.

Puppán, Dániel 2002. Environmental Evaluation of Biofuels. *Periodica Polytechnica Series Social and Management Sciences*. 10/1, 95–116.

Purchase, Iain F.H. 2005. What Determines the Acceptability of Genetically Modified Food that Can Improve Human Nutrition? *Toxicology and Applied Pharmacology*. 207, 819–827.

Raats, Monique M. / Shepherd, Richard / Sparks, Paul 1995. Including Moral Dimensions of Choice Within the Structure of the Theory of Planned Behavior. *Journal of Applied Social Psychology*. 25/6, 484–494.

Ramesohl, Stephan / Merten, Frank 2006. Energy System Aspects of Hydrogen as an Alternative Fuel in Transport. *Energy Policy*. 34/11, 1251–1259.

Rawwas, Mohammed Y. A. / Patzer, Gordon L. / Klassen, Michael L. 1995. Consumer Ethics in Cross-Cultural Settings. *European Journal of Marketing*. 29/7, 62–78.

Rimal, Arbindra P. / Moon, Wanki / Balasubramanian, Siva K. 2005. Agro-Biotechnology and Organic Food Purchase in the United Kingdom. *British Food Journal*. 107/2, 84–97.

Rimal, Arbindra P. / Moon, Wanki / Balasubramanian, Siva K. 2006. Perceived Risks of Agro-Biotechnology and Organic Food Purchases in the United States. *Journal of Food Distribution Research*. 37/2, 70–79.

Roddy, Gerardine / Cowan, Cathal A. / Hutchinson, George 1996. Consumer Attitudes and Behavior to Organic Foods in Ireland. *Journal of International Consumer Marketing*. 9/2, 41–63.

Roff, Robin J. 2008. No Alternative? The Politics and History of non-GMO Certification. *Agriculture and Human Values*. Retrieved May 3rd, 2009 from: <http://dx.doi.org/10.1007/s10460-008-9166-5>.

Rogers, Everett M. 2003. *Diffusion of Innovations*. New York, NY: Free Press.

Roselli, Luigi / Seccia, Antonio / Stasi, Antonio 2006. Atteggiamento dei Consumatori nei Confronti dell'Evoluzione del Sistema Agro-Alimentare: L'Introduzione di Alimenti Geneticamente Modificati. *Economia Agro-Alimentare*. 10/1, 127–150.

Rozin, Paul 2005. The Meaning of "Natural". *Psychological Science*. 16/8, 652–658.

Russi, Daniela 2008. An Integrated Assessment of a Large-Scale Biodiesel Production in Italy: Killing Several Birds with One Stone. *Energy Policy*. 36/3, 1169–1180.

Saba, Anna / Messina, Federico 2003. Attitudes towards Organic Food and Risk/Benefit Perception Associated with Pesticides. *Food Quality and Preference*. 14, 637–645.

Schifferstein, Hendrik N.J. / Oude Ophuis, Peter A.M. 1998. Health-Related Determinants of Organic Food Consumption in the Netherlands. *Food Quality and Preference*. 9/3, 119–133.

Schulte, Inga / Hart, David / Van der Vorst, Rita 2004. Issues Affecting the Acceptance of Hydrogen Fuel. *International Journal of Hydrogen Energy*. 29/7, 677–685.

Schwartz, Shalom H. 1992. Universals in the Content and Structure of Values: Theory and Empirical Tests in 20 Countries. In Zanna, Mark P. (ed.) *Advances in Experimental Social Psychology*. New York: Academic Press. 25, 1–65.

Schwartz, Shalom H. / Bilsky, Wolfgang 1987. Toward a Universal Psychological Structure of Human Values. *Journal of Personality and Social Psychology*. 53/3, 550–562.

Schwartz, Shalom H. / Bilsky, Wolfgang 1990. Toward a Theory of the Universal Content and Structure of Values: Extensions and

Cross-Cultural Replications. *Journal of Personality and Social Psychology*. 58/5, 878–891.

Shaw, Deirdre / Clarke, Ian 1998. Culture, Consumption and Choice: Towards a Conceptual Relationship. *Journal of Consumer Studies and Home Economics*. 22/3, 163–168.

Shaw, Deirdre / Clarke, Ian 1999. Belief Formation in Ethical Consumer Groups: An Exploratory Study. *Marketing Intelligence & Planning*. 17/2, 109–120.

Shaw, Deirdre / Newholm, Terry 2002. Voluntary Simplicity and the Ethics of Consumption. *Psychology and Marketing*. 19/2, 167–185.

Shaw, Deirdre / Shiu, Edward 2002a. An Assessment of Ethical Obligation and Self-Identity in Ethical Consumer Decision-Making: A Structural Equation Modelling Approach. *International Journal of Consumer Studies*. 26/4, 286–293.

Shaw, Deirdre / Shiu, Edward 2002b. The Role of Ethical Obligation and Self-Identity in Ethical Consumer Choice. *International Journal of Consumer Studies*. 26/2, 109–116.

Shaw, Deirdre / Shiu, Edward 2003. Ethics in Consumer Choice: A Multivariate Modelling Approach. *European Journal of Marketing*. 37/10, 1485–1518.

Shaw, Deirdre / Shiu, Edward / Clarke, Ian 2000. The Contribution of Ethical Obligation and Self-Identity to the Theory of Planned Behavior: An Exploration of Ethical Consumers. *Journal of Marketing Management*. 16/8, 879–894.

Shaw, Deirdre S. / Grehan, Emma / Shiu, Edward / Hassan, Louise / Thomson, Jennifer 2005. An Exploration of Values in Ethical Consumer Decision Making. *Journal of Consumer Behaviour*. 4/3, 185–200.

Sheth, Jagdish N. 1970. Are There Differences in Dissonance Reduction Behavior Between Students and Housewives? *Journal of Marketing Research*. 7/2, 243–245.

Siegrist, Michael 1998. Belief in Gene Technology: The Influence of Environmental Attitudes and Gender. *Personal individual Differences*. 24/6, 861–866.

Siegrist, Michael 1999. A Causal Model Explaining the Perception and Acceptance of Gene Technology. *Journal of Applied Social Psychology*. 29/10, 2093–2106.

Siegrist, Michael 2003. Perception of Gene Technology and Food Risks: Results of a Survey in Switzerland. *Journal of Risk Research*. 6/1, 45–60.

Soler, Francisco / Gil, José M. / Sanchez, Mercedes 2002. Consumer's Acceptability of Organic Food in Spain: Results from an Experimental Action Market. *British Food Journal*. 104/8, 643–653.

Solomon, Barry D. / Banerjee, Abhijit 2006. A Global Survey of Hydrogen Energy Research, Development and Policy. *Energy Policy*. 34/7, 781–792.

Solomon, Michael R. / Bamossy, Gary / Askegaard, Soren 1999. *Consumer Behaviour: A European Perspective*. Harlow: Prentice Hall.

Sparks, Paul / Guthrie, Carol A. 1998. Self-Identity and the Theory of Planned Behavior: A Useful Addition or an Unhelpful Artifice? *Journal of Applied Social Psychology*. 28/15, 1393–1410.

Sparks, Paul / Shepherd, Richard 1992. Self-identity and the Theory of Planned Behavior: Assessing the Role of Identification with "Green Consumerism". *Social Psychology Quarterly*. 55/4, 388–399.

Sparks, Paul / Shepherd, Richard 2002. The Role of Moral Judgments within Expectancy-Value-Based Attitude-Behavior Models. *Ethics & Behavior*. 12/4, 299–321.

Spence, Alexa / Townsend, Ellen 2006. Examining Consumer Behavior toward Genetically Modified (GM) Food in Britain. *Risk Analysis*. 26/3, 657–670.

Springer, Antje / Tsioumanis, Asterios / Papastefanou, Georgios / Mattas, Konstadinos 2005. Using Value Priorities to Explain Differences in Attitudes Towards Genetically Modified Food. Paper presented at the XI Congress of European Association of Agricultural Economists: *The Future of Rural Europe in the Global Agri-Food System*. Copenhagen, DK, August 24–27. Retrieved November 29th, 2008, from: <http://ageconsearch.umn.edu/bitstream/24657/1/pp05sp01.pdf>.

Srinivasan, Sunderasan 2009. The Food v. Fuel Debate: A Nuanced View of Incentive Structures. *Renewable Energy*. 34/4, 950–954.

Steenhaut, Sarah / Van Kenhove, Patrick 2006. An Empirical Investigation of the Relationships among a Consumer's Personal Values, Ethical Ideology and Ethical Beliefs. *Journal of Business Ethics*. 64, 137–155.

Suslow, Trevor V. / Thomas, Bruce R. / Bradford, Kent J. 2002. Biotechnology Provides New Tools for Plant Breeding. Agricultural Biotechnology in California Series-Publication 8043. University of California – Division of Agriculture and Natural Resources. Retrieved November 29th, 2008 from: <http://anrcatalog. ucdavis.edu>.

Sutton, Stephen 1998. Predictive and Explaining Intentions and Behavior: How Well We Do? *Journal of Applied Social Psychology*, 28/15, 1317–1338.

Tarkiainen, Anssi / Sundqvist, Sanna 2005. Subjective Norms, Attitudes and Intentions of Finnish Consumers in Buying Organic Food. *British Food Journal*. 107/11, 808–822.

Thøgersen, John 1999. The Ethical Consumer. Moral Norms and Packaging Choice. *Journal of Consumer Policy*. 22/4, 439–460.

Thøgersen, John 2002. Direct Experience and the Strength of the Personal Norm – Behavior Relationship. *Psychology & Marketing*. 19/10, 881–893.

Thøgersen, John / Ölander, Folke 2002. Human Values and the Emergence of a Sustainable Consumption Pattern. *Journal of Economic Psychology*. 23, 605–630.

Thøgersen, John / Ölander, Folke 2006. The Dynamic Interaction of Personal Norms and Environment-Friendly Buying Behavior: A Panel Study. *Journal of Applied Social Psychology*. 36/7, 1758–1780.

Thompson, Gary D. 1998. Consumer Demand for Organic Foods: What We Know and What We Need to Know. *American Journal of Agricultural Economics*. 80/5, 1113–1118.

Thompson, Paul B. 2008. The Agricultural Ethics of Biofuels: A First Look. *Journal of Agricultural and Environmental Ethics*. 21/2, 183–198.

Titus, Philip A. / Bradford, Jeffrey L. 1996, Reflections on Consumer Sophistication and Its Impact on Ethical Business Practice. *Journal of Consumer Affairs*. 30/1, 170–194.

Trucker, Mark / Whaley, Sherrie R. / Sharp, Jeff S. 2006. Consumer Perception of Food-Related Risks, *International Journal of Food Science and Technology*. 41, 135–146.

Tsioumanis, Asterios / Mattas, Konstadinos / Papastefanou, Georgios / Springer, Antje 2004. Potential on Using Cultural Syndromes for Explaining Differences in Attitudes in Northern and Southern EU Countries. Paper presented at the 84th EAAAE Seminar: *Food Safety in a Dynamic World*. Zeist, NL, February 8–11. Retrieved November 29th, 2008, from: <http://ageconsearch.umn.edu/bitstream/24984/1/sp04ts01.pdf>.

United States Department of Agriculture 2008. Retrieved February, 20th 2008 from: <http://www.usda.gov/wps/portal/!ut/p/_s.7_0_A/7_0_1OB?contentidonly=true&navid=AGRICULTURE&contentid=BiotechnologyGlosary.xml>.

Ureña, Felix / Bernaéu, Rodolofo / Olmeda, Miguel 2008. Women, Men and Organic Food: Differences in their Attitudes and Willingness to Pay. A Spanish Case Study. *International Journal of Consumer Studies*. 32, 18–26.

Van Dijk, Heleen / Houghton, Julie / Van Kleef, Ellen / Van Der Lans, Ivo / Rowe, Gene / Frewer, Lynn 2008. Consumer Responses to Communication about Food Risk Management. *Appetite*. 50, 340–352.

Van Ruijven, Bas / Hari, Lakshmikanth / Van Vuuren, Detlef P. / De Vries, Bert 2008. The Potential Role of Hydrogen Energy in India and Western Europe. *Energy Policy*. 36/5, 1649–1665.

Varaldo, Riccardo / Guido, Gianluigi 1997. Il Consumatore Come Prodotto: "Customer Satisfaction" Come Qualità del Cliente. *Micro e Macro Marketing*, 6/1, 9–40.

Verplanken, Bas / Holland, Rob W. 2002. Motivated Decision Making: Effects of Activation and Self-Centrality of Values on Choices and Behavior. *Journal of Personality and Social Psychology*. 82/3, 434–447.

Vitell, Scott J. 2003. Consumer Ethics Research: Review, Synthesis and Suggestions for the Future. *Journal of Business Ethics*. 43/1-2, 33–47.

Vitell, Scott J. / Muncy, James 1992. Consumer Ethics: An Empirical Investigation of Factors Influencing Ethical Judgments of the Final Consumer. *Journal of Business Ethics*. 11/8, 585–597.

Vitell, Scott J. / Muncy, James 2005. The Muncy-Vitell Consumer Ethics Scale: A Modification and Application. *Journal of Business Ethics*. 62/3, 267–275.

Vitell, Scott J. / Paolillo, Joseph G. P. / Singh, Jatinder J.2006. The Role of Money and Religiosity in Determining Consumers' Ethical Beliefs. *Journal of Business Ethics*. 64/2, 117–124.

Von Alvensleben, Reimar 2001. Beliefs Associated with Food Production Methods. In Frewer, Lynn J. / Risvik, Einar / Schifferstein, Hendrik N.J. (eds) *Food, People and Society – A European Perspective of Consumer's Food Choices*. Berlin Heidelberg New York: Springer-Verlag, 381–400.

Wang, Qingbin / Halbrendt, Catherine / Webb, Shwu-Eng H. 1997. Consumer Demand for Organic food in China: Evidence from Survey Data. *Agricultural Production and Nutrition: Proceedings of an International Conference*. Boston, MA, March 19–21, 1997.

Wheale, Peter / Hinton, David 2007. Ethical Consumers in Search of Markets. *Business Strategy and the Environment*. 16/4, 302–315.

Wier, Mette / Calverley, Carmen 2002. Market Potential for Organic Foods in Europe. *British Food Journal*. 104/1, 45–62.

Wier, Mette / Smed, Sinne 2000. Modeling Consumption of Organic Foods. In Alföldi Thomas / Lockeretz William / Niggli, Urs (eds.) *IFOAM 2000 – The World Grows Organic*. Zurich: Vdf Hochschulverlag, 550.

Williams, Pamela R.D. / Hammitt, James K. 2000. A Comparison of Organic and Conventional Fresh Produce Buyers in the Boston Area. *Risk Analysis*. 20/5, 735–746.

Williams, Pamela R.D. / Hammitt, James K. 2001. Perceived Risks of Conventional and Organic Produce: Pesticides, Pathogens, and Natural Toxins. *Risk Analysis*. 21/2, 319–330.

Woese, Katrin / Lange, Dirk / Boess, Christian / Bogl, Klaus W. 1997. A Comparison of Organically and Conventionally Grown Foods – Results of a Review of the Relevant Literature. *Journal of Science Food and Agriculture*. 74, 281–293.

Wolf, Marianne M. 2002. An Analysis of the Impact of Price on Consumer Interest in Organic Grapes and a Profile of Organic Purchasers. Paper presented at the *American Agricultural Economics Association Annual Meeting*. Long Beach, California, July 28–31.

Yang, Hong / Zhou, Yuan / Liu, Junguo 2009. Land and Water Requirements of Biofuel and Implications for Food Supply and the Environment in China. *Energy Policy*. 37/5, 1876–1885.

Yavas, Ugur 1994. Research Note: Students as Subjects in Advertising and Marketing Research. *International Marketing Review*. 11/4, 35–43.

Yeniyurt, Sengun / Townsend, Janell D. 2003. Does Culture Explain Acceptance of New Products in a Country? *International Marketing Review*. 20/4, 377–396.

Zanoli, Raffaele / Naspetti, Simona 2002. Consumer Motivations in the Purchase of Organic Food: A Means-End Approach. *British Food Journal*. 104/8, 643–653.

Zechendorf, Bernhard 1998. Agricultural Biotechnology: Why do Europeans Have Difficulty Accepting it? *Agbioforum*. 1/1, 8–13.

Zerta, Martin / Schmidt, Patrick R. / Stiller, Christoph / Landinger, Hubert 2008. Alternative World Energy Outlook (AWEO) and the Role of Hydrogen in a Changing Energy Landscape. *International Journal of Hydrogen Energy*. 33/12, 3021–3025.

Zhao, Jimin / Melaina, Marc W. 2006. Transition to Hydrogen-Based Transportation in China: Lessons Learned from Alternative Fuel Vehicle Programs in the United States and China. *Energy Policy*. 34/11, 1299–1309.

Authors

Gianluigi Guido is Full Professor of Marketing and Place Marketing at the University of Salento at Lecce, Italy, and Professor of Marketing Research at the LUISS University of Rome. He holds a M.I.B.A. from the U.S. International University, San Diego (California), and a Ph.D. in Management Sciences from the University of Cambridge (UK). He has been Professor of Marketing at the University of Padua, and at the *Sapienza* University of Rome; visiting researcher at the Dept of Marketing of the University of Florida at Gainesville; and Fulbright fellow at the Dept of Psychology of the University of Stanford (California). Prof. Guido has published eleven books on consumer behaviour and marketing strategies, and more than 100 articles in major referred journals, among which: *International Journal of Market Research, Journal of Business & Industrial Marketing*, *Journal of Consumer Marketing*, *Journal of Current Issues and Research in Advertising*, *Journal of Economic Psychology*, *Journal of Euromarketing*, *Journal of Product and Brand Management*, *Journal of Satisfaction, Dissatisfaction & Complaining Behavior*, *Long Range Planning*, *Perception & Motor Skills*, *Recherche et Applications en Marketing*, *Research Policy*, and others. His publications in English include: *Computer-Mediated Relationships and Trust* (L.L. Brennan and V.E. Johnson, eds), New York 2008; *Content and Media Factors in Advertising* (P. Neijens *et al.*, eds), Amsterdam 2004; *The Salience of Marketing Stimuli*, Boston, MA 2001; *Readings in Global Marketing* (M.R. Czinkota and I.A. Ronkainen, eds), London 1995. He has been listed in the last twelve editions of Marquis *Who's Who in the World* and in the last six editions of Marquis *Who's Who in Finance & Industry*.

Carolina Buffa holds a M.Sc. in Economics and Business Management from the LUISS University of Rome (Italy), and an International Diploma in Marketing from the University of California at Berkeley. She has worked as Marketing Research Assistant for Comcast Spotlight, San Francisco, and for the European Market Division of Lottomatica, Rome. She is currently Assistant Brand Manager for Unilever Italia, Rome. Her primary interests fall within the areas of marketing and brand management.

Laura Franceschini holds a First Class Degree in Industrial Economics, with specialization in Marketing, from the LUISS University of Rome (Italy), and a Specialization Degree in Management Innovation and Internationalization from the *Sapienza* University of Rome. She has worked as Assistant Brand Manager ("Home Care" Area) for the Colgate Palmolive corporation. She is currently working as Assistant to the Company Director in the Administrative Department of INA-Assitalia, an insurance company in the Generali Group.

R. Christian Maloumby-Baka is Policy and a Social Dialogue Analyst and International Communication Consultant. He has worked in the field of international cooperation and development since 1995, both in Europe and Africa. By working for several NGOs and international institutions, such as the OECD, he has promoted, managed, and strengthened partnerships with bilateral and multilateral institutions. He received a Master degree in Economics and Business from the University of Salento in Lecce (Italy) and attended the MPhil in Economic Governance course at the *Institut d'Etudes Politiques* in Paris (France). He has been lecturer in International Human Resources Management at the LUMSA University in Rome.

Alessandro M. Peluso holds a Ph.D. in Economic and Quantitative Methods for Market Analysis from the University of Salento at Lecce (Italy). He is currently Research Associate in Marketing at the LUISS University of Rome, and Research Assistant in Marketing at the University of Salento, Lecce. His primary research interests fall within the areas of consumer satisfaction, innovation adoption and diffusion, and source-perceived credibility in advertising contexts. He has published articles in international referred journals, including: *International Journal of Market Research*, *Journal of Current Issues and Research in Advertising*, *Journal of Product and Brand Management*, and *Research Policy*.

Giovanni Pino holds a *Laurea* degree in Economics and Business Administration, with a major in Marketing, at the Dept of Economics of the University of Salento at Lecce (Italy). He is currently doctoral student in Intelligent Management at the International School of Advanced Studies of the "Istituto Superiore per la Formazione Universitaria" (ISUFI) in Lecce. His research activity concentrates on consumer behavior and health care management, with a special focus on food consumption.

M. Irene Prete holds a Ph.D. in Economic and Quantitative Methods for Market Analysis at the University of Salento at Lecce, Italy, and a Science Master (*Master Recherche*) in Management at the Ecole Doctorale Economie, Gestion et Espace at the University of Paris XII (France). She is actually Research Associate in Marketing and Place Marketing at the University of Salento at Lecce. Her research interests are in the field of political marketing, virtual communities and new media marketing. She has published in several national and international academic journals. Her recent publications include *Computer-Mediated Relationships and Trust* (L.L. Brennan and V.E. Johnson, eds), New York 2008.

Piermario Tedeschi is an expert in digital marketing and new media. He works as Online Marketing Strategist for Social Media and Web consultant for affiliation programs and mobile marketing. After three years as Assistant Researcher in Marketing Research and Advertising at the LUISS University of Rome, he moved to London where he started working as a Project Manager in 77Agency, a new media agency. He holds a Bachelor's Degree in Marketing Management from the LUISS University of Rome and a Master's Degree in Management, Innovation and Internationalization from the *Sapienza* University of Rome. He is also registered in the Trainee Professional Accountant Roll in Italy.

Index

Luk Bouckaert / Hendrik Opdebeeck /
Laszlo Zsolnai (eds)

Frugality

Rebalancing Material and Spiritual Values in Economic Life

Oxford, Bern, Berlin, Bruxelles, Frankfurt am Main, New York, Wien, 2008.
XIV, 322 pp., 1 coloured ill.
Frontiers of Business Ethics. Vol. 4
Edited by Laszlo Zsolnai
ISBN 978-3-03911-131-2 pb.
sFr. 96.– / € 65.90 / €** 67.80 / € 61.60 / £ 40.– / US-$ 95.95*

* incluye el IVA – valido para Alemania ** incluye el IVA – valido para Austria

This book examines frugality as an ideal and an 'art de vivre' which implies a low level of material consumption and a simple lifestyle, to open the mind for spiritual goods as inner freedom, social peace and justice or the quest for God or 'ultimate reality'. By rational choice we can develop a more frugal and sufficient way of life, but material temptations can always overwrite our ecological, social and ethical considerations. But the spiritual case for frugality is strong enough. Spiritually based frugal practices may lead to rational outcomes such as reducing ecological destruction, social disintegration and the exploitation of future generations.

Contents: Luk Bouckaert/Hendrik Opdebeeck/Laszlo Zsolnai: Why Frugality – Luk Bouckaert: Rational versus Spiritual Concepts of Frugality – Rafael Esteban: Frugality and the Body – Francis Kadaplackal: How the Idea 'Created Co-Creator' Can Contribute to the Nurturing of Frugality in Economic Life – Laurie Michaelis: Quaker Simplicity – Dirk Geldof: Overconsumption – Ronald Commers/Wim Vandekerckhove: Frugality and the Moral Economy of Late Capitalism – Knut Ims/Ove Jacobsen: Consumerism and Frugality: Contradictory Principles in Economics? – Hendrik Opdebeeck: The Urgency of a Frugality-based Economics – Herman E. Daly: Frugality First – Ronald Jeurissen/Bert van de Ven: Frugal Marketing: Can Selling Less Make Business Sense? – Robert H. Frank: The Progressive Consumption Tax – Laszlo Zsolnai: Buddhist Economic Strategy.

Kolibri 360 GmbH
Ferdinand-Jühlke-Straße 7
99095 Erfurt, Deutschland
produktsicherheit@kolibri360.de